P9-BYL-547

Other books by Richard Morris

Light

The End of the World

The Fate of the Universe

Evolution and Human Nature

The Word and Beyond (coauthor)

Dismantling the Universe

Time's Arrows

Scientific Attitudes Toward Time

Richard Morris

Simon and Schuster
New York

Published by Simon and Schuster
A Division of Simon & Schuster, Inc.
Simon & Schuster Building
Rockefeller Center
1230 Avenue of the Americas
New York, New York 10020
SIMON AND SCHUSTER and colophon are registered
trademarks of Simon & Schuster, Inc.
Designed by Barbara Marks
Manufactured in the United States of America
2 3 4 5 6 7 8 9 10
Library of Congress Cataloging in Publication Data
Morris, Richard, date.
Time's arrows.

Bibliography: p.
Includes index.
1. Time. I. Title.
QB209.M67 1984 115 84-16146
ISBN: 0-671-50158-5

Contents

CHAPTER 1

What Is Time?

SOME QUESTIONS SEEM TOO trivial to merit serious consideration, but often these are the very questions that become most puzzling when we examine them in detail. Things are often not as simple as they seem. Suddenly we perceive hidden complexities, and the apparently simple question becomes baffling.

The question "What is time?" falls into this category. We all know what time is. It is what a clock measures. Nothing could be simpler than that. But when we begin to look at the question in detail, we discover that the topic of time is complicated indeed, that "What is time?" is, in reality, a whole series of related questions. For example, what is this thing that we call the "flow of time"? Does time always progress at the same rate? Is it like a flowing river, or is it the moment we call "now" that moves from the present into the future? Is it conceivable that the flow of time could be stopped or reversed? Did time have a beginning? If so, how did it come into existence? Will time have an end? What was happening before the universe was created? Is time nothing more than a succession of events, or does it possess a kind of independent reality? Just what *is* this thing that a clock measures?

Approximately sixteen hundred years ago, St. Augustine made a particularly apt observation about this puzzling subject. "What, then, is time?" he asked. "If no one asks me, I know what it is. If I wish to explain it to him who asks, I do not know." Of course, we should probably not take this statement too literally. When philosophers say they do not know the answer to a question, we can generally expect that the profession of ignorance will be followed by a lengthy analysis. St. Augustine was no exception. After making this disclaimer, he dwelt on the subject of time at some length, and eventually concluded that time was something that resided in the soul.

I do not propose to discuss Augustine's analysis of time in detail. Nor will I dwell, at any great length, upon the opinions of the other philosophers who have attempted to come to grips with the subject of time. I am more interested in treating "What is time?" as a scientific question. It seems to me that we must discover what science can tell us about time before we can know what the relevant philosophical questions are. If the subject is approached in any other way, one can only expect to engage in needless intellectual effort. For example, it would be useless to philosophize about the beginning of time while remaining ignorant of what physics and cosmology have to say on the subject.

Since the beginning of the scientific revolution of the sixteenth century, science has appropriated many of the traditional concerns of philosophy. For example, to the Greeks of the classical era, attempting to understand the nature of the physical world was a philosophical activity. Plato discussed cosmology at great length in his dialogue *Timaeus,* and Aristotle discussed natural phenomena in such books as his *Physics* and *Meteorology.* As late as the eighteenth century, the German philosopher Immanuel Kant was speculating about the nature of space. Today, all of these things are part of the subject matter of such sciences as physics, chemistry, and cosmology.

Although there are aspects of the topic of time that are

still discussed by philosophers, the question "What is time?" is also being appropriated by science. The development within physics of such disciplines as mechanics, thermodynamics, and electrodynamics has made it possible to ask meaningful questions about time and to obtain at least tentative answers. In the field of cosmology, the development of the big bang theory of the origin of the universe has allowed us to offer scientific speculations about how time began. And, of course, Einstein's theories of relativity have made it possible to perform experiments and to engage in theoretical analyses which have a direct bearing upon the nature of space and time.

The twentieth century has been the great age of discovery in physics. Consequently I was tempted to begin by discussing some of the developments that took place shortly after 1900, for example, Einstein's special theory of relativity, which was propounded in 1905. I concluded, however, that this wasn't the best approach. After all, modern ideas about time did not begin with Einstein. On the contrary, they evolved slowly, over a period of centuries. If one gains an understanding of how this evolution of ideas took place and knows something about the various paths taken by those who thought about the subject, it is easier to see why time is viewed as it is today.

There are significant differences between a topic like time and such scientific subjects as relativity or the theory of subatomic particles. Relativity was discovered in the first decade of our century. The discovery that atoms are not the basic constituents of matter, that they are made up of yet smaller particles, is even more recent. On the other hand, time is a subject that humanity has pondered for millennia, perhaps since the very beginning of civilization.

Over the centuries, there has been a continual interaction of philosophical outlooks, social trends, and scientific ideas about time. Consequently, our understanding of modern views on time will be enhanced if we can see how and why these interactions took place.

In most past ages, concepts of time have been dramatically different from what they are today. It was not only the philosophers and intellectuals who understood time in a different manner than we do; ideas about time have tended to be regarded as part of the commonsense outlook on reality. But what was "common sense" two thousand or so years ago bears little relation to what is thought to be "common sense" today. For example, most ancient civilizations did not share the modern view of time as a linear continuum that stretches into the indefinite future. Ancient peoples believed that time was cyclic in character—that historical events followed cyclic patterns, and that these patterns were somehow reflected in the nature of time itself.

In civilization after civilization, we encounter myths that tell us that the world will eventually be destroyed, and that it will then be recreated so that a new cycle can begin. The world will experience patterns of events, these myths tell us, that are fated to be repeated over and over again. Sometimes these cycles were complex indeed. For example, in ancient Greece, it was commonly believed that time exhibited a pattern called the Great Year. The world was destined to be destroyed by ice during a Great Winter, and by fire during the Great Summer. After each cataclysm, the world would be created anew, and humanity would again progress through ages of gold, silver, bronze, and iron.

Some, such as the followers of the philosopher Pythagoras, the Stoics, and certain Neoplatonic philosophers, believed in the doctrine of eternal recurrence. They thought that living human beings were destined to be born again during future cycles, and that the same (or similar) events would happen over and over again. It was believed by some that there would be another Athens, another Socrates, another trial, and another drinking of the hemlock. Naturally, not everyone accepted these ideas. But even Aristotle, who repudiated the doctrine of eternal recurrence, believed that history followed cyclic patterns, and that there was a sense in which time could be thought of as being circular.

We, on the other hand, habitually think of time as something that stretches in a straight line into the past and future. We do not believe that time is circular, that there will be another age like classical Greece, or that Socrates (or someone like him) will live again. It is true that there are traces of the cyclical idea in our thinking. For example, there are those who perceive cycles in history, and those who espouse cyclical theories of the stock market. However, for the most part, we tend to view time as a continuum.

This manner of thinking is part of the heritage of Christianity. The early Christian writers stressed the importance of individual historical events that would not be repeated. History, they said, did not move in cycles. On the contrary, there had been a Creation at a particular point in time. Christ had died on the Cross but once, and had been resurrected from the dead on but one occasion. Finally, at some point in the future, God's plan would be completed, and He would—once and for all time—bring the world to an end.

The linear concept of time has had profound effects on Western thought. Without it, it would be difficult to conceive of the idea of progress, or to speak of cosmic or biological evolution. After all, if the same patterns were destined to repeat themselves over and over again, real progress would be impossible. Furthermore, if there was evolution, it would have to be a kind of evolution that would repeat itself endlessly. A cosmos that repeats itself is fundamentally different from one that was created at a given point in time and has evolved to its present state.

The shift from a cyclical to a linear outlook on time is not the only profound change that has taken place. For example, today we think of time as an abstract quantity that can be broken down into hours, minutes, and seconds. But throughout most of history, time was not viewed in this way. Until the end of the medieval era, it was something that revealed itself in the rhythms of nature and of everyday life. When the day was divided into hours, these hours were not of constant length. There were twelve hours of daylight and twelve hours of darkness. As a result, daytime and nighttime hours varied in length throughout the year, and were equal to one another only at the spring and fall equinoxes. But, of course, it really didn't make very much difference how long the hours were. The unit of labor was not the hour, but the day. Work ordinarily began at dawn and ended at sunset. Only the monks in the monasteries had any need to know the hour of the day, so that they would know when to perform their devotions.

But as clocks began to proliferate during the fourteenth century, time began to become something that one could see ticking away. Gradually, it became possible to think about time in ways that would have been inconceivable in earlier eras. It was during this period that hours began to be divided into minutes and seconds, and that workers in the cities first began to start and stop work at definite times. Time was slowly transformed into an abstract quantity that existed in its own right.

The newfound ability to conceive of time in an abstract way was one of the factors that precipitated the scientific revolution of the sixteenth century. Modern physics can be said to have begun with Galileo's studies of motion, but Galileo would not have made any of his discoveries in this field if he had not realized that time could be used as a tool in the analysis of the motion of physical bodies. It would not be stretching the truth to say that it was Galileo who discovered that time was a physical quantity.

Galileo began by working out the laws that governed the motion of falling bodies. He was the first to realize that the acceleration of these bodies was a function of time, that the same increment of velocity was added to their speed of fall every second. This was a profound discovery indeed. Galileo's predecessors had generally not given much thought to the function of time. They had tended to think either that the velocity of a falling object must be constant, or that velocity was proportional to distance traveled, not time.

Galileo's discoveries will be discussed in detail in Chapter 3. At the moment it is enough to point out that his discovery of time as a tool for the analysis of the physical world set science on a course it is still following today. There is hardly a field in physics that does not concern itself with time. Physicists speak of the billions of years that were required for the evolution of the universe, and of the tiny fractions of a second that represent the lifetimes of unstable subnuclear particles. The mathematical tool used in all fields of physics, calculus, was originally invented because scientists needed a way to study the physical evolution of systems in time.

In the chapters that follow I plan to look in detail at the evolution of ideas about time. Without these changes in outlook, modern science would not have been possible. Indeed, science developed in Western culture, and it was only Western culture that embraced the idea of abstract, linear time.

As scientists have tried to understand the nature of time, they have uncovered new puzzles as often as they have found answers. The more they have learned about time, the more baffling the subject has become. For example, physicists have discovered four different "arrows" that can be used to define the direction of time. Yet no one is quite sure how the different arrows of time are related to one another, or to what can be considered a fifth arrow, the psychological perception of time. It has been discovered that under certain circumstances, it is possible to speculate about time that may go backward, or about objects that may travel backward in time. Physicists have discovered that the times experienced by different observers cannot always be synchronized with one another. And yet there is a cosmic time that characterizes the universe as a whole. Scientists have discovered evidence that there exists a single, rare subatomic interaction that behaves differently with respect to time than any other. But they do not know what significance, if any, this fact has. Finally, they have speculated about the beginning of time, and have realized that no one really knows whether time is a linear phenomenon that had a beginning, or whether it might not be cyclical in character after all.

I don't mean to imply that nothing is understood about time. On the contrary, the realization that a subject is puzzling is often a sign that a certain degree of understanding has been attained. I, for one, believe that he who considers "What is time?" a baffling question has a much deeper understanding of the subject than he who thinks that time is merely "what a clock measures."

CHAPTER 2

Cyclical Time and Linear Time

THE ANCIENT BABYLONIANS were skilled astronomers who maintained records of their observations over a period of centuries. By 1800 B.C., they were already compiling star catalogs and recording planetary movements. By the middle of the eighth century B.C., they were keeping dated records of celestial observations and applying mathematical techniques that were as sophisticated as those used by Western astronomers in the time of Copernicus. Even after the Babylonians were conquered by Alexander the Great in 331 B.C., they continued to refine their methods of observation, with striking results. During the latter part of the fourth century B.C., the Babylonian astronomer Kidinnu calculated the motion of the sun with an accuracy that was not exceeded until the twentieth century.

The Babylonians achieved these results without telescopes, without accurate clocks, and without any of the measuring instruments that are found in modern observatories. They relied, instead, on records of astronomical observations that had been compiled over periods of centuries. Their results were precise because observations had been made continuously. If individual measurements contained inaccuracies, the effects were minimal. Over periods of hundreds of years, the errors tended to cancel one another out.

The Babylonians did not study events in the heavens out of intellectual curiosity, however. They propounded no theories to explain the motions they saw in the night sky. They were concerned only with keeping records and with predicting eclipses, conjunctions, and retrogradations. They had no interest in explaining why celestial objects behaved the way they did; they wanted only to know what had happened in the past, and to develop techniques for forecasting the events that would be observed in the future.

The Babylonians believed that the heavens were divine. Each of the planets was identified with one of the Babylonian gods, and it was believed that by observing their motions, one could discern the gods' intentions. The Babylonian astronomers were not so much scientists as they were priests who looked for, and attempted to interpret, celestial omens.

The arts of divination were extensively practiced in Babylonia and in other ancient Mesopotamian civilizations. From time immemorial, priests had been attempting to predict future events by examining the viscera of animals, by observing the patterns formed when oil was poured into water, and by interpreting such ominous events as the birth of malformed animals. They did not doubt that the fortunes of kings and of states could be learned by communicating with supernatural beings and forces.

The Mesopotamian kings sometimes distrusted the predictions of their diviners, but when they did, their skepticism was most often the result of suspicions concerning the diviners' professional honesty; they did not doubt that a king had a duty to receive oracles from the gods, and to conduct the affairs of state accordingly.

Thus Mesopotamia became the birthplace of astrology. At first, the diviners attempted only to interpret celestial portents. But as Babylonian astronomical methods became more refined, the prediction of the movements of planets and predictions of eclipses assumed a greater importance. By the time of the Chaldean dynasty (625 to 539 B.C.), astrological divination had become important indeed. The Chaldeans believed that celestial events could be used to forecast not only the fortunes of kings and of nations, but the fate of each individual.

The Chaldeans may not have cast horoscopes, however. This type of astrology may have developed only later, during the fifth century B.C., after Babylonia had been conquered by the Persians. But the Chaldean astronomers did develop the idea that terrestrial events followed a cyclic pattern. The stars and the planets, after all, moved in cycles. Hence it seemed only reasonable to believe that similar cycles could be observed on earth.

Sometime around 290 or 270 B.C., according to ancient writers, the Babylonian priest Berossos migrated to the Greek island of Cos, where he lectured on Babylonian philosophy. According to the Roman philosopher Seneca, who lived approximately three hundred years later, Berossos expounded a doctrine concerning the Great Year. Apparently Berossos taught that the world would periodically be destroyed, and then created anew, at periodic intervals, when all the stars came together in the constellation of the Crab. Each re-creation would mark the beginning of a new Great Year, during which terrestrial events would parallel those of the Great Year that had just ended. According to this doctrine, terrestrial events exhibited cyclic patterns that paralleled those which could be seen in the heavens.

We don't know how accurately Seneca's account reflects beliefs that were common during the Chaldean dynasty. In particular, it is hard to believe that the Chaldeans thought that all the stars would eventually come together in one section of the sky. They were certainly expert enough in astronomy to know that only the planets changed their relative positions.

However, it is very likely that the Chaldeans believed in some sort of Great Year. References to the idea often appear in the writings of the Greeks of the classical era, and it is probable that the Greeks took the idea from the Chaldeans, who were the source of much of their astronomical knowledge.

During classical times, the trade routes from Mesopotamia led to Greek ports on the Mediterranean. Knowledge as well as goods passed over these routes, and as a result the Greeks were able to familiarize themselves with Babylonian culture. The Greeks made use of Babylonian astronomical records, and Greek writers make scattered references to Chaldean astrology. It seems reasonable to infer that the stimulus for Greek speculation about the Great Year must have been contact with Mesopotamian civilization.

In his dialogue *Timaeus,* Plato gives a definition of the Great Year that sounds a little more reasonable than Seneca's. According to Plato, a Great Year would come to an end when all of the planets moved back into the positions that they had occupied at one time long ago. A new Great Year would then begin, and end when the planets once again returned to their original positions. Plato did not speculate about the length of a Great Year. However, later commentators set it at 36,000 years.

Plato does not suggest that the end of a Great Year is marked by any cosmic cataclysm. He does not even suggest that a Great Year has any definite beginning or end. After all, the positions of the planets are the same as what they were one Great Year ago at every moment. The planets are analagous to a clock that keeps perfect time.

Of course, there was no such thing as a mechanical clock in Plato's day. However, it would not be amiss to liken his view of planetary motion to the movement of the hands of a clock. Every twelve hours, the same positions recur. There is no definite beginning or ending, because a twelve-hour cycle can start anywhere.

In Plato, the Great Year does not have the same significance that it did to the priest Berossos and to those who, like Seneca, popularized his doctrines. Plato was less interested in propounding a theory of cyclical events than he was in the nature of time itself. In Plato's view, the movement of the planets *was* time. Time came into being with the heavens; time could not exist apart from them. The heavens were "a moving image of eternity."

In Plato's dialogue *Timaeus,* we are told:

> So time came into being with the heavens in order that, having come into being together, they should also be dissolved together if ever they are dissolved; and it was made as like as possible to eternity, which was its model.

And:

> Only a very few men are aware of the periods of [the planets]. . . . They are indeed virtually unaware that their wandering movements are time at all, so bewildering are they in number and so amazing in intricacy. None the less it is perfectly possible to perceive that the perfect temporal number and the perfect year [Great Year] are complete when all eight orbits have reached their total of revolutions relative to each other, measured by the regularly moving orbit of the Same.

The idea that time is inseparable from the periodic motions that are observed in the heavens is not unique to Plato, but runs through all of Greek thought. Time is always associated with circular movement. Consequently, time itself is often thought of as a circle. When a Great Year is completed, not only the planets but time itself returns to its starting point.

The concept of circular time was sometimes associated with the doctrine of eternal recurrence. The Pythagoreans, the Stoics, and the Neoplatonist philosophers who elaborated upon Plato's doctrines after his death all believed that the same individuals would be born again, and that the same or similar events would take place over and over. Although this belief was not universally held, even those who argued against it tended to think of time as cyclical, and to see cyclic patterns in the world around them.

Aristotle maintained that all things did not return upon themselves in a uniform manner. Specifically, he thought that the coming-to-be of human individuals constituted a linear, not a cyclic, sequence. "For though your coming-to-be presupposes your father's," Aristotle noted in *On Coming-to-Be and Passing-Away* (often referred to by its Latin title, *De generatione et corruptione*), "his coming-to-be does not presuppose yours." Yet even Aristotle spoke of human opinions as coming to be in cycles that happened infinitely often. Furthermore, he too believed that time was circular. In his *Physics,* he tells us that "if one and the same motion sometimes recurs, it will be one and the same time." In another passage that can be found in the same book, he elaborates as follows:

> This also is why time is thought to be movement of the sphere, viz. because the other movements are measured by this, and time by this movement. This also explains the common saying that human affairs form a circle, and that there is a circle in all other things that have a natural movement and coming into being and passing away. This is because all other things are discriminated by time, and end and begin as though conforming to a cycle; for even time itself is thought to be a circle.

In his *Problems,* Aristotle goes so far as to suggest that we might be living before as well as after the time of Troy. "If then human life is a circle," he says, "and a circle has neither beginning nor end, we should not be 'prior' to those who lived in the time of Troy nor they 'prior' to us by being nearer to the beginning." The point is that since a circle has no beginning or end, "before" and "after" have no absolute meaning.

Such a manner of thinking is likely to seem strange to the modern mind. "How can time be circular," one is tempted to ask, "if the same events do not repeat themselves endlessly?" Perhaps one should liken the Greek concept of time to our idea of "time of day." We speak of going to work "at the same time" every day, or going to bed or getting up at a definite "time." In order to understand the Greek view, we must imagine extending this concept to events that unfolded over periods of thousands of years. To the Greeks, the entire cosmos, not just the alternations of night and day, exhibited cyclic patterns.

Some of the Greeks did, of course, think of time as circular in the sense that exactly the same events would be repeated endlessly. This was one of the doctrines of Stoicism, a school of philosophy that was founded by Zeno of Citium (no relation to Zeno the Eleatic, author of the famous paradoxes) in the third century B.C.

The Stoics drew the natural conclusion that circular time implied a rigid determinism. Since the events that were destined to take place in the future would only be repetitions of those that had happened in the past, human beings were powerless to influence the course of events. According to the Stoic view, only the human will was free. Though one could not alter the course that his life was destined to take, it was possible to cultivate a virtuous will, and an attitude of resignation. According to the Stoic philosopher and Roman emperor Marcus Aurelius, it was possible for the virtuous soul to "reach out into eternity, embracing and comprehending the great cyclic renewals of creation, and thereby perceiving that future generations will have nothing new to witness."

The Greeks and Romans were not the only people in ancient times to think of time as cyclic. Indian philosophy of the Vedic period (around 1500 to 600 B.C.) conceived of cycles within cycles. The smallest was an age, which measured about 360 human years; the longest corresponded to the lives of the gods, which were thought to be of the order of 300 trillion years. But time would not come to an end, not even after those trillions of years had passed. The gods themselves would die and be reborn, and the cosmic cycles of creation and destruction would go on forever.

The ancient Chinese conceived of the cosmos as exhibiting a cyclic interplay between yin and yang. After Chinese astronomy had advanced to the point that it was possible to determine planetary periods, a world cycle of 23,639 years was calculated. The Aztecs, on the other hand, were somewhat more parsimonious in their thoughts about time. They had a fifty-two-year cycle. It was thought that at the end of a cycle, the world was in danger of being destroyed. Sometimes this danger came to pass, and sometimes it didn't. When it did, the world moved into a new age, or "Sun."

The Mayas believed in cyclical time and cyclical catastrophes also. In fact, in 1698 members of the Mayan Itzan tribe fled from a group of invading Spaniards because they believed that the time of catastrophe had come. It wasn't by chance that the Spaniards decided to conquer the Itza in that year. The tribe had been visited by missionaries some eighty years before. At that time, the fathers had learned of the date calculated for the end of the current cycle.

Belief in world cycles and in the periodic destruction and recreation of the cosmos was also part of Norse mythology. According to the collection of myths known as the *Prose Edda,* heaven and earth would be destroyed at *Ragnarok* ("Twilight of the Gods"). But this would not represent the end of time, for the world was destined to be created again. According to these myths, there would be a new heaven and earth, new gods and a new human race.

Time has always been ordered by observations of cyclic events such as the rising and setting of the sun, the phases of the moon, and the alternation of the seasons. As soon as man began to observe the stars, he became aware that there were periodic motions in the heavens also. When the idea of "time" was conceived, nothing seemed more natural than to relate it to these periodic events. Since the idea of cyclic time appears in so many cultures, one must conclude that it must have seemed almost unnatural to think of it as a linear phenomenon.

The idea of linear time was introduced into Western thought by Judaism and by Christianity. Judaism was unique among the religions of the ancient world in that it placed an emphasis on unique historical events that were supposed to have taken place at particular points in time. The Exodus from Egypt, for example, was something that had happened only once. It was a specific event that had an important religious significance. Similarly, God had made his promises to Abraham on a specific occasion, and He had given the Law to Moses at a particular time. History, in ancient Judaism, was an arena in which God's purposes were fulfilled. It would have been practically blasphemous to suggest that historical events were repeated in endless cycles.

The concept of cyclical time was not unknown to the ancient Jews. We read in the book of Ecclesiastes, "The thing that hath been, it is that which shall be; and that which is done is that which shall be done; and there is no new thing under the sun." However, references to such ideas are encountered infrequently in the Old Testament. The emphasis is on the working out of God's purpose in a linear time that began with the Creation.

Christianity emphasized the concept of linear time to an even greater degree. The central event in Christian doctrine was the death and resurrection of Christ. To imply that similar events had taken place on numerous occasions in different cosmic cycles would be to destroy the meaning of the redemption. "So Christ was once offered to bear the sins of many," says St. Paul in his Epistle to the Hebrews. "For Christ also hath once suffered for sins," we are admonished in 1 Peter 3:18.

According to Christian doctrine, the world had been created at a definite time, and it would end on an unknown date. The world was not eternal, as the Greeks had believed; on the contrary, it had had a beginning, and it would have an end.

The nature of time was a preoccupation of a number of the early church fathers, particularly of St. Augustine, who became bishop of Hippo in North Africa in 396 A.D. In Augustine's time, the cyclical conception of time was well known in the Roman world, and Neoplatonic philosophers were expounding the doctrine of eternal recurrence. Since such ideas could not be reconciled with Christian doctrine, they obviously had to be combated. In his *City of God,* Augustine charged that the doctrine of cyclical time was a concept promulgated by "deceiving and deceived sages." Not only was it a foolish doctrine, it was also impious. Like St. Paul, Augustine took pains to emphasize that "once Christ died for our sins; and, rising from the dead, He dieth no more."

Augustine also repudiated the doctrines associated with astrology. Although he sometimes seems to have been bothered more by astrological determinism than by the long-standing association between astrology and ideas about cyclical time and the Great Year, he was surely aware that these connections existed. It was most likely not by chance that astrology had experienced a revival among the Romans (the Greeks had paid little attention to it) at about the same time that Stoic and Neoplatonic ideas became popular.

Although the idea of cyclical time was vehemently opposed by the church, it did not die out completely. Speculation about cosmic cycles continued well into medieval times. It must have been discussed quite frequently during the thirteenth century, for example. In 1277, Étienne Tempier, bishop of Paris, cited it as the sixth of 219 opinions that, he said, should be condemned as heretical.

But the heresy did not die out. During the Renaissance, the Great Year was a much-discussed topic. As late as the nineteenth century, the German philosopher Friedrich Nietzsche advocated the doctrine of eternal recurrence and cyclical time. If the universe was finite, and if time was infinite, Nietzsche argued, then it was inevitable that the same events would happen again and again. Although Nietzsche didn't argue, as the Greeks had, that time itself was circular, he did believe that events would eventually be repeated.

The early Christian fathers developed the concept of linear time in some detail. For example, according to Augustine, history could be divided into six stages in analogy with the six days of Creation. The first age began with the Creation and ended with the Deluge. The second began with Noah and ended with Abraham. The third stretched from Abraham to David, the fourth concluded with the Babylonian Captivity, and the fifth with the birth of Christ. The sixth, and final, age had begun with the birth of Christ; it would continue until the Last Judgment, when God would bring time to an end.

This scheme became more or less orthodox during the Middle Ages. To be sure, there were variations. For example, some theologians divided history into four, not six, periods, corresponding to the Babylonian, Persian, Macedonian, and Roman empires. However, few of them doubted that they were living in the last age of the world. According to a generally accepted view, the birth of Christ marked the beginning of the world's "old age."

No one knew what date had been appointed for the Second Coming. However, all agreed that it would not take place in the indefinitely remote future; humanity was living out its last days. Consequently, medieval culture was not preoccupied with ideas of social or intellectual progress. The amount of time that remained was limited, and in any case the world was nothing more than a transitory resting place. Man's goal was not to seek betterment in this world, but to prepare for eternity.

It may be that such ideas had something to do with the extravagant tenor that sometimes characterized life in medieval times. The men and women who lived in that era seem to have enjoyed going to extremes to a greater extent than people in most other ages. The wealthy often dressed extravagantly, gave extravagant feasts, and indulged in extravagant pleasures. They did not hesitate to indulge themselves in excesses of emotion. Although they adhered to doctrines that were otherworldly in principle, they behaved as though they had to make the fullest possible use of the little time that remained.

Even the common people indulged themselves when they could. When wandering preachers spoke of the Last

Judgment, of hell, or of the Passion, they would sometimes elicit such bitter sobbing that the sermon would have to be suspended until the weeping had stopped. Nor did expressions of emotion necessarily have to be connected with religion. On one occasion, the citizens of Mons, in southwestern Belgium, bought a brigand in order to experience the pleasure of seeing him quartered. According to a contemporary observer, this entertainment caused the onlookers to rejoice "more than if a new holy body had risen from the dead."

And then there were the saints, who would sometimes mix ashes in their food or test their virtue by sleeping beside young women. During medieval times, religious relics were adored beyond all reason, whether they were likely to be genuine or not. It has been said that medieval Europe contained enough splinters of the True Cross to build a fleet of ships if they could all have been collected and put together. Sometimes corpses provided valuable relics too. After St. Thomas Aquinas died in their monastery, some monks decapitated the body and boiled the head in order to preserve it as a relic. On one occasion, King Charles VI of France distributed some of the bones of his ancestor St. Louis at a feast. And sometimes the relic-seekers were not even willing to wait until a saint was dead. Around the year 1000, some Umbrian peasants attempted to murder St. Romuald, the hermit, in order to possess his bones.

Whether this extravagance was the result of the conviction that the world was in its last days or not, it is clear that the people of the medieval era did not view time the way we do. Although the linear time of Christianity had replaced the cyclical time of antiquity, time was not something that existed in abundance. During the Middle Ages, it was believed that the total span of time from the Creation to the Second Coming was extremely limited. Some estimated it at six thousand years; others professed not to know exactly how much had been allotted. However, it was clear to all that whatever the length of time God had provided, worldly time paled before eternity.

The medieval sense of time differed from ours in another respect also. During the Middle Ages, individuals did not have the awareness of time that we possess today. They did not wear watches that divided time into hours, minutes, and seconds. They did not make appointments for specific times. They began work at sunrise, not at a time that was the same on every working day throughout the year. During the greater part of the Middle Ages, most individuals never knew what time it was.

The mechanical clock was not invented until the thirteenth century. At first, it was little more than a curiosity. Those who needed to know the time of day in a more or less approximate way, such as the monks, who were required to say devotions at certain hours, continued to rely upon such devices as water clocks and sundials.

In the fourteenth century, the invention of the escapement, a device that controlled the movements of a mechanical clock, made greater accuracy possible. At last it was possible to construct clocks that would keep time to an accuracy of about fifteen or thirty minutes per day. Public clocks began to become more common, and around 1345, the hour was divided into minutes and seconds for the first time.

But it was only at the very end of the medieval period that the majority of the population were even likely to know what a clock was. Life was regulated not by the hours and minutes of abstract time, but by the rhythms of everyday life and of the seasons. Work began at sunrise and ended at sunset. If this made the working day longer at the beginning of summer than it was at any other time of the year, no one complained. After all, that was something that God, in His inscrutable wisdom, had ordained. Although the seasonal agricultural rhythms and the festivals of the church were a constant reminder of the passage of a year, few among the peasant population ever knew what year it was. Possibly there were numerous members of the aristocracy who did not know either. This was an age in which only clerics knew how to read and to write.

CHAPTER 3

Abstract Time

THE INVENTION AND DEVELOPMENT of the mechanical clock was one of the most significant technological developments that has taken place within Western society. As the social historian Lewis Mumford has pointed out, it was the clock that dissociated time from natural rhythms and "helped to create the belief in an independent world of mathematically measurable sequences" that was so essential to the development of science. It was the clock that increased the importance of worldly time, and that lessened the medieval preoccupation with eternity. And it was the clock that made time into an abstract entity that could be contemplated in its own right as a sequence of hours, minutes, and seconds.

Although clocks were being constructed by the second half of the thirteenth century, at first they were not very accurate devices. As often as not, they would gain or lose hours over the course of a day. The problem was that no one had devised a way to regulate the speed of a clock's movements. Horologists understood well enough that a clock could be driven by weights, and they knew how to make toothed wheels. But, as Richard the Englishman commented in 1271, the clockmakers could not "quite perfect their work."

The problem of regulating a clock's movement was

solved sometime near the end of the century, when the verge and foliot escapement was invented. This was a mechanism that controlled the rate at which a clock ran. It was made up of a verge, which checked the motion of a toothed wheel, and a foliot, a swinging rod that performed the same function as that of the balance wheel in a modern watch or clock. It is the operation of the escapement that causes a clock to "tick."

Many of the first clocks that used the escapement principle were constructed during the early fourteenth century for use in monasteries. Unlike modern clocks, they did not have dials. The clocks were simply devices that caused bells to be struck on the hour. Although it is hard to determine exactly how accurate they were, it is unlikely that they kept time to an accuracy of better than a half hour per day.

Somewhat later during the fourteenth century, clocks with dials began to appear. Most of these were public clocks, which were placed in churches and cathedrals. At first they only had hour hands. It wasn't until the sixteenth century that it was thought necessary to add a minute hand to show the subdivisions of the hour.

The accuracy of the clocks that were constructed during the fourteenth century was not great. Apparently it was not thought necessary to measure time with any great precision. In any case, medieval clockmakers found it easier to add wheels and gears and make their mechanisms more complicated than to find ways of regulating the movement of the escapement. Nevertheless, many of the fourteenth-century clocks were quite remarkable. Some incorporated moving calendars and had astronomical sections that indicated the positions of the sun, the moon, and the planets. Naturally, these complex devices had a tendency to break down. As a result, it was generally thought necessary to employ a "governor" to oversee the operation of a clock and to reset the time periodically. Sometimes even this did not prove to be sufficient. In 1387, King John of Aragon appointed two men to strike bells to indicate the time because his clock failed to do this correctly. At about the same time, the clock in the royal palace in Paris proved so unreliable that the following rhyme was made up about it: *L'horloge du palais, elle va comme il lui plaît,* "The clock of the palace runs as it pleases."

Although the early clocks were inaccurate, expensive devices that required constant tending, clockmaking became one of the major industries of the fourteenth century. There were undoubtedly a number of reasons for this. Civic pride was one. Cities vied with one another to acquire complicated clocks for their cathedrals. The fact that many clocks showed the movements of the planetary bodies was also important, since astrology was experiencing a revival at this time. Knowledge of the positions of the planets was thought essential for the success of various kinds of enterprises.

The proliferation of clocks had a considerable effect on late medieval and early Renaissance culture. The passage of time suddenly became something of which one was very much aware. One could "see" time simply by watching the hand of a clock move around a dial. Time was no longer simply a sequence of experiences; it had become something that was measured by the passing hours and their subdivisions.

As the Middle Ages ended and the Renaissance began, outlooks on time changed rapidly. One way to illustrate these changes would be to compare references to time in the work of the Italian poets Dante and Petrarch. Dante, who was born in 1265, exhibits a characteristic medieval outlook. On the other hand, Petrarch, who was born thirty-nine years later, and who is often considered the first modern poet, views the subject of time quite differently. Where Dante shares the medieval preoccupation with eternity, Petrarch tends to view time as a commodity that can be saved or squandered. He is conscious of the passing of time in a way that Dante is not.

Dante speaks of time in a number of places in his *Divine Comedy*. For example, in Canto XXXI of *Paradiso,* he speaks of having come "to the eternal from time." In Canto XXVII, he emphasizes the insignificance of worldly time by again comparing it to eternity. Time, says Dante, has its roots in the *primum mobile* (in Ptolemaic astronomy, the outermost of the celestial spheres); but on earth we see only time's passing leaves. In Canto XXXII of *Purgatorio,* Beatrice tells Dante that he will be a dweller on earth for only a "brief" time, and adds, allegorically, that he will be "forever a citizen of that Rome in which Christ is a Roman."

Petrarch, on the other hand, is obsessed with the passage of time on earth. He complains of the brevity of life, of its "swiftness, haste, tumbling course," and laments "time's irrecoverability, the flower of life soon wasted, the fugitive beauty of a rosy face, the frantic flight of unreturning youth, the trickeries of stealthy age." Life, says Petrarch, is a race toward death in which "all are propelled with a uniform motion and driven along with no variation in the rate of progress."

Both Dante and Petrarch agree that life is brief, but we should not be misled by this apparent similarity in outlook. Dante seems not to be concerned about the passing of time. In Petrarch, it becomes an obsession. Dante mentions the briefness of life only to emphasize the relative importance of eternity. On the other hand, when Petrarch speaks of life's brevity, he is speaking of it in terms of a kind of time that can be quantified.

The discovery that time possessed a terrifying reality seems to have had quite an impact on Petrarch, for during the latter part of his life he made heroic efforts to conserve all of the time that remained to him. It was folly, he said in numerous letters written to his contemporaries, to waste even one day. Life, he said, "continually flies and consumes itself . . . every day carries you toward old age. While you are looking around, while you delay, suddenly gray hair is upon you." Perhaps it is not without significance that the figure of Father Time had its origins in some drawings that were made to illustrate the poems in Petrarch's *Trionfi* ("Triumphs").

But changes in the prevailing outlooks on time were not confined to poetry. Effects can also be seen in the economic sphere. For example, during the Middle Ages, one of the reasons that were given for the prohibition of usury was that the charging of interest on money was tantamount to selling time. Time was supposed to belong to God alone; thus the usurer was selling something that did not belong to him.

At the beginning of the Renaissance, usury was still prohibited. However, the growth of capitalism caused this prohibition to be gradually relaxed. The early capitalists realized that time was something that could be measured and used, and as they became a more important factor in society, their outlooks began to be adopted by society as a whole. Merchants realized that the lengths of sea voyages and land journeys affected their profits, as did such factors as the rise and fall of prices in time and the working hours of their employees (laborers and craftsmen were commonly employed by merchants in these days). Bells, and then clocks, began to be used to chime the hours of commercial transactions and of hours of labor. Greater attention was paid to the organization of time in the workplace. Though the unit of labor was still the day, rather than the hour, work was now supposed to begin and end at fixed times. For the first time in human history, clock time was used to regulate the lives of individuals.

Around the end of the sixteenth century, the new concept of time as something that could actually be measured and used began to have an impact upon science. It was at about this time that the Italian scientist Galileo realized that if one was to have any hope of devising a theory that would explain the motion of objects through the air, it was necessary to explain the role that was played by time.

There are some scientific discoveries that seem so obvious that it is difficult to see why they were important, and why they were not made much earlier than they were. To us, it seems intuitively obvious, for example, that falling bodies are accelerated, that their velocity increases with time. Galileo's discovery that falling bodies behave in this manner seems anything but a great insight. We find it hard to think that anyone could ever have believed anything else.

But in Galileo's day this fact was anything but obvious.

Some denied that falling bodies experienced any acceleration at all. This view was even shared by Galileo at one time; when he wrote on the subject as a young man, he maintained that the apparent acceleration of a falling object was an illusion.

To better understand the problems that Galileo faced, it might be a good idea to take a brief look at some of the attempts of his predecessors to solve the puzzle of motion. In particular, it might be well to consider some of the theories of Aristotle, for Aristotelian ideas were still generally accepted during the latter part of the sixteenth century.

Aristotle distinguished between several different types of motion, including augmentation, alteration, natural motion, and violent motion. The last two were types of motion that were characteristic of moving bodies. The term "natural motion" described the tendency of a heavy object to fall as well as to the tendency of fire or smoke to rise. According to Aristotle, objects fell because they were seeking their natural place near the center of the universe. Furthermore, Aristotle said, the velocity at which an object fell was proportional to its weight.

Aristotle believed that violent motion, the motion of an object that was thrown in a horizontal direction or up in the air, was a different kind of phenomenon. Unlike natural motion, violent motion required a motive force. Since, according to Aristotelian doctrines, nothing could move unless there was a mover, it followed that an object could move horizontally, or upward, only when a force was exerted on it continuously. An arrow or a rock flew through the air because something was pushing it along. The moment that the force ceased to be applied, the violent motion would cease, and natural motion toward the center of the earth would take its place.

Aristotle suggested that it was the air itself that provided this motive force. When an arrow left a bowstring, for example, the air that it displaced would rush in behind it, and this disturbance would propel the arrow forward. Aristotle also believed, paradoxically, that the resistance of the air would simultaneously cause the arrow to slow and eventually fall to the ground. The medium through which an object traveled not only provided a motive force, but also the resistance that would cause motion to cease.

Practically all of Aristotle's conclusions about motion were mistaken. Nevertheless, he did succeed in knitting his ideas together in a logical, reasonably consistent system. This system seemed so impressive that thinking about the problems of motion continued to be based on Aristotelian ideas up to the time of Galileo. Philosophers and scientists would sometimes question one or another of Aristotle's conclusions, but it seems not to have occurred to any of them that the entire Aristotelian theory should be thrown overboard.

It would be misleading to suggest, however, that Galileo's predecessors had made no progress toward gaining a better understanding of the motion of objects through the air. This was not exactly the case. After Aristotle's works were translated into Latin during the twelfth century, medieval scholars did attempt to elaborate upon the theories that they discovered in the philosopher's *Physics*. These speculations led to some real accomplishments.

For example, there was the fourteenth-century scholar William of Heytesbury. Much of Heytesbury's work seems outlandish by modern standards. One problem that he considered involved the rotation of a body whose outermost parts were continually being destroyed or corrupted while the inner parts expanded or were rarefied. It seems not to have occurred to Heytesbury that there was really little point in expending so much effort contemplating the behavior of a combustible piece of doughnut dough which rises as its rotating outer parts burn away.* Science in the Middle Ages was something of a scholastic pursuit that had little or no connection with experiment or with practical applications.

However, Heytesbury and his colleagues at Oxford did anticipate Galileo by studying the problem of accelerated motion. Unfortunately, they did not see the connection be-

* My example, not Heytesbury's.

tween acceleration and the motion of freely falling bodies; they seem to have been concerned only with abstract speculation. However, Galileo would have found it much easier to develop an understanding of the behavior of moving bodies if the Oxford scholars' theories had been widely known in his time.

The greatest achievement of medieval physics was the development of the theory of impetus, which did provide Galileo with a starting point for his own speculations. The impetus theory, which was perfected by the fourteenth-century Parisian philosopher Jean Buridan, was based on some observations about motion that had been made by the Greek writer John Philoponus in the sixth century A.D.

Philoponus was skeptical about Aristotle's assertion that it was contact with the air that was responsible for violent motion. He realized that if air did provide the motive force, then it ought to be possible to move a stone by agitating the air behind it. Since this was obviously not the case, Philoponus looked for an alternative explanation. He concluded that there must be a motive force that resided within the stone itself. If such a force could be imparted to a stone by the object that moved it (for example, the thrower's hand), motion could be easily explained. According to Philoponus, the impressed force would cause the stone to continue to move until air resistance or a collision with another body counteracted it.

Buridan developed these ideas further, and gave the name "impetus" to Philoponus' hypothetical force. He hypothesized that impetus depended upon both speed and the quantity of matter in a body, and he reasoned that after a body had lost contact with its mover, impetus would not diminish when no other forces were acting. Thus, under ideal conditions, the body would continue to travel in a straight line at uniform speed.

Buridan also used the theory to explain the behavior of falling bodies. Unlike many of Galileo's contemporaries, Buridan realized that an object did accelerate as it fell. Following Aristotle, he assumed that a falling body first acquired a velocity that was proportional to its weight. Then, as it fell, the weight imparted a quantity of impetus, which increased as the fall continued. The increasing impetus caused the object to move faster and faster until it struck the ground.

Although impetus theory had a certain number of successes, there was also one significant defect: It could not easily be checked against experiment. The theory could not be used to calculate how long it should take an object to fall from a given height, for example. Nor did it predict how far a projectile would travel when it was fired or thrown in a horizontal direction.

Neither Buridan nor his contemporaries considered this to be a defect, however. They did not concern themselves with making numerical predictions, or with checking theoretical predictions against experiment. Like Aristotle, they were interested only in explaining the causes of motion. They wanted to know *why* objects behaved as they did rather than *how*.

The distinction between these two types of explanation is a significant one. It illustrates the difference between medieval and modern science. The modern scientist wants to know what happens in nature, and to find mathematical laws that describe the phenomena he observes. He is not interested in the question "Why?" He does not ask why atoms are made up of positively charged nuclei that are surrounded by negatively charged electrons, but he does ask how atoms are bound together in molecules, and about the physical mechanisms that cause atoms to emit light. He doesn't ask why electricity and magnetism should be related to one another, but how; and he demands mathematical equations that will allow him to calculate, for example, the exact quantity of electrical current that will be induced when a magnet is moved at a certain speed through a loop of wire.

Galileo must be considered to be the first modern physicist, because he was the first to emphasize the overriding importance of the question "How?" in scientific explanation.

In his writings, Galileo emphasized that science had to concern itself with the "sensible world" (or, as we say today, the "real world"), not the world of abstract argument. Today, this idea seems almost a platitude. But in Galileo's time it represented an entirely new kind of outlook. The idea was so novel that even the great French philosopher and mathematician René Descartes failed to understand it. Writing during the early part of the seventeenth century, Descartes argued against Galileo's theories of motion on the grounds that the causes of motion had not been explained.

It is impossible to say how Galileo got his revolutionary idea. When Galileo began pondering the problems of motion during the latter part of the sixteenth century, there was no such thing as experimental physics. Rather than observe nature, scientists continued to expound the theories of Aristotle and the two-century-old theory of impetus as well. It never occurred to them to investigate how moving objects actually behaved.

Perhaps it was Galileo's natural rebelliousness that prompted him to investigate nature, and that led eventually to his discovery of the importance of time in physics. Like many great innovators, Galileo delighted in flouting convention, at least in the intellectual arena. For example, when he wrote about his discoveries, Galileo rarely passed up a chance to ridicule Aristotelian ideas.

The beginning of Galileo's search for an understanding of the puzzle of motion seems to have begun sometime around the year 1590. In 1589, when Galileo was twenty-five, he took a position as professor of mathematics at the University of Pisa. He taught at Pisa until 1592, when he accepted the more highly paid chair of mathematics at the University of Padua. While he was at Pisa, Galileo began writing a treatise on mechanics* called *De motu*. This work was never published during Galileo's lifetime. However, manuscript

* Mechanics is the branch of physics that deals with energy and forces and their effects on bodies.

versions of it still exist. These make it clear that Galileo was already questioning the tenets of Aristotelian physics when he was still a young man.

Galileo maintained that, contrary to Aristotelian doctrine, objects did not fall with velocities that were proportional to their weight. On the contrary, Galileo maintained, all bodies composed of the same material fell at the same velocity. A ten-pound lead ball would not fall ten times as fast as one that weighed one pound; on the contrary, if the two balls were dropped at the same time from the same height, they would reach the ground together. Galileo was later to generalize this assertion and to maintain that all objects, of any material, would fall at the same rate in the absence of air resistance. However, in his Latin treatise *De motu,* he concerned himself only with bodies that had the same material composition.

It is unlikely, by the way, that Galileo ever tried to prove his case by dropping weights from the Leaning Tower of Pisa. This famous story seems to have been invented by Vincenzio Viviani, a pupil of Galileo's who wrote a rather romantic biography of his master after the latter's death. According to Viviani, the experiment was performed in the presence of Galileo's students and some of the other professors at the University of Pisa. But there are no contemporary accounts of the demonstration, and scholars doubt that it ever took place.

On the other hand, we do know that experiments of this type were performed in 1612 by a professor at Pisa who was attempting to demonstrate the validity of Aristotle's theories of motion. He found, as he expected, that the heavier object did reach the ground before the lighter one. The difference in time of descent was small, but it could be observed.

Today, no one would find this result to be very surprising. The lighter object will be impeded more by air resistance. But, of course, this does not demonstrate the correctness of Aristotelian theory, as Galileo himself pointed out in his last book, *Two New Sciences*:

> Aristotle says, "A hundred-pound iron ball falling from the height of a hundred braccia [the braccio was a measure of length equal to 58.4 centimeters, or about 23 inches] hits the ground before one of just one pound has descended a single braccio." You find on making the experiment that the larger anticipates the smaller by two inches; that is, when the larger one strikes the ground, the other is two inches behind it. And now you want to hide, behind those two inches, the ninety-nine braccia of Aristotle, and speaking only of my tiny error, remain silent about his enormous one.

Although Galileo understood that the Aristotelian description of free fall was incorrect at the time that he wrote *De motu,* he had not yet devised a correct theory of his own. He had not yet realized that acceleration was a fundamental characteristic of the motion of falling objects. He was aware that acceleration was observed, but he tried to explain it away.

Galileo gave two, somewhat contradictory, explanations for the observed acceleration of falling bodies. At one point, he said that it was an illusion, caused by tricks of perspective. At another, he attempted to use the impetus theory to show that the acceleration of a falling body was only a temporary phenomenon that lasted, at most, a few seconds. If an object was allowed to fall for long enough a time, he claimed, it would attain a velocity that would thereafter remain constant.

According to Galileo, motionless objects gained impetus from the surfaces on which they rested. A ball on a table, for example, would receive from the table an impetus that just balanced its weight. If the ball was removed from the table and allowed to fall, this impetus had to be overcome before it could attain its characteristic velocity. As the impetus withered away, it would move faster and faster. But after the impetus was gone, there would be no further acceleration.

Galileo never published *De motu.* Most likely, the reason he did not was that his observations of falling bodies failed to confirm his theory. Around the time that he was writing this treatise, Galileo began to observe the behavior of balls that he rolled down inclined planes. Apparently, he soon obtained results that were at odds with his theoretical arguments.

Galileo realized that he could not learn much by watching objects fall. Falling bodies accelerated so rapidly that it was difficult to tell what was really happening. An object that is dropped from a height of ten feet, for example, reaches the ground in four-fifths of a second; if the height is increased to a hundred feet, the time of descent is still only two and one-half seconds. Galileo had no way to measure such short times with any great accuracy. Nor could he determine what the velocity of a falling body was at any given point of its descent.

However, Galileo saw that he could "dilute" the effects of gravity by allowing balls to roll down planes. If a ball was allowed to descend to the ground along a gentle slope, its time of fall would be much longer. For example, if a ball is placed at a ten-foot height, it will take more than twenty-two seconds to roll down a plane that is inclined at a 2° angle. It can still be viewed as "falling" under the influence of gravity. However, it does this in slow motion.

As he performed his experiments, Galileo began to realize that acceleration was always exhibited by falling bodies. Furthermore, he was able to work out a mathematical formula that related time to the distance that an object fell. Specifically, he found that distance was proportional to the square of the time. If an object descended, say, one foot in the first second, then one would find that it had fallen four feet after two seconds, nine feet at the end of the third second, and so on.

Descriptions of this sort sound cumbersome when expressed in words. The relationship between distance and time can be expressed in a simpler manner if one uses the following mathematical formula:

$$d \propto t^2,$$

where $\propto$ is a mathematical symbol meaning "is proportional to." Note that this formula relates distance to *elapsed* time. It describes how far an object has fallen at the end of (for example) three seconds, not how far it falls during the third second alone.

Galileo was not the first to realize that time and distance fallen were somehow related; nothing could be more obvious than that. He was not even the first to work out a mathematical formula relating elapsed time to distance traveled in accelerated motion. This had been done more than two hundred years before by the group of medieval Oxford physicists (who failed to relate their result to the problem of free fall, however). However, Galileo was the first to time physical events. As far as we know, at the time that he performed his experiments, it had never occurred to anyone to attempt to measure times of free fall.

Once the relationship between distance and time had been established, the obvious next step was to find the relationship between time and velocity. But Galileo did not take this step at once. He seems to have assumed, as most of the scholars of his day did, that velocity and distance fallen were proportional to one another.

As a matter of fact, they are not. Velocity is proportional to the square root of the distance that an object has fallen. When it has fallen through a distance of four feet, it will be moving only twice, not four times, as fast as it was when it had gone only one foot. But Galileo cannot be blamed for failing to realize this at once. When he assumed that velocity and distance were proportional, he was simply going along with accepted belief.

It took quite some time before he began to question this assumption. In the autumn of 1604, Galileo happened to mention his discovery concerning distance and time to his Venetian friend Fra Paolo Sarpi. Sarpi asked for a mathematical demonstration of these results. Galileo duly wrote one out. It is not known whether he ever sent it to Sarpi. However, it is preserved among his manuscripts.

Galileo's demonstration was fallacious. He erroneously thought that he had demonstrated the distance-time-squared proportionality by making the assumption that velocity and distance were proportional. But velocity is actually proportional to *time*. Although a falling body does not move at four times the velocity after it has gone four feet, it does travel four times as fast at the end of four seconds.

Again, all of this may become somewhat clearer if the behavior of a falling body is expressed in mathematical terms, rather than in words. This time, I will write down equations, rather than expressions of proportionality. The relationship between velocity and time is

$$v = at.$$

Velocity is equal to the product of acceleration and distance. Similarly, the relationship between distance and time can be expressed by the equation

$$d = \frac{1}{2}at^2.$$

This exact formula was found by Galileo shortly after he realized that distance was proportional to the square of the time.

Galileo eventually discovered the correct expression for velocity also. No one knows exactly when he found it, or how he discovered it. We only know that it was not easy for him to reach this conclusion. For example, he seems to have thought, for a while, that velocity was proportional to both distance and time. It may have taken him until around 1615 to see clearly what the correct relationships were.

We should not find it surprising that it took Galileo so long to find the solution to the problem. After all, the velocity of an accelerating body was a quantity that was not easy to define. When an object moves at a constant speed, there are no problems. If it is traveling at a velocity of ten feet per second, then it will move ten feet in one second, twenty feet in two seconds, and so on. However, in the case of accelerating objects, Galileo had to deal not with constant velocity but with *instantaneous velocity* that was constantly changing at every moment in time.

When an object is accelerated, the velocity is never the same from one moment to the next; on the contrary, it is continually increasing. Instantaneous velocity cannot even be measured directly. The best that one can do is to find the average speed over some small time period.

Determining that instantaneous velocity was proportional to one quantity but not to another was not so easy a task. But Galileo finally solved the problem. After more than twenty-five years of thought on the subject, he finally knew what happened when objects fell.

Once he had worked out that problem, determining the motion of a body moving in any arbitrary direction was easy. For example, if one knows how accelerated bodies behave, it is not difficult to determine what will happen to a stone that is thrown in a horizontal direction. Obviously, it will follow a path that curves downward. Its inertia will carry it along in a horizontal direction as gravity causes it to accelerate in a vertical direction. When the horizontal and vertical motions are combined, a geometrical curve known as a *parabola* will result.

Similarly, one can calculate what will happen to a cannonball that is fired up at an angle in order to attain greater distance. As in the case of the stone, the vertical and horizontal motions can be considered separately. Since the cannonball is accelerated by gravity only in the vertical direction, its inertia will again carry it along at a constant horizontal speed. The upward component of its motion will gradually decrease as gravity causes the cannonball to decelerate. Once it begins to fall downward, it will accelerate in the same manner as the stone.

The result is again a parabola, but one that has two "legs" instead of one. Naturally this trajectory is perfectly symmetrical, provided that we assume (as Galileo did) that the effects of air resistance are so small that they can be disregarded. In such a case, what goes up must come down in exactly the same manner.

A parabola, incidentally, is also the curve that is described by the large cables that support suspension bridges. If one mentally turns such a bridge upside down, the result is a curve that describes the path of a projectile. If we cut the curve in two so that we are left only with the half that represents downward falling motion, we have a curve that represents the path of the horizontally thrown stone.

It was probably no accident that Galileo wrote about the trajectories of projectiles. In his day, as in ours, it was realized that the military could benefit by making use of advances in science and in technology. Galileo moonlighted by teaching mathematics to military officers who wanted to learn how to increase the effectiveness of artillery pieces. He also invented, manufactured, and sold an instrument that was used for measuring the elevation of cannons.

It may be that science benefited from his preoccupation with military problems. His analysis of projectile motion could be applied to any body that moved about in the vicinity of the earth's surface. After all, any object that is thrown, heaved, propelled, or shot in any direction behaves in exactly the same way; its momentum carries it along horizontally, while gravity creates a vertical acceleration. After Galileo had made his discoveries in mechanics, all that remained to be done was to solve problems concerning circular motion, and to investigate the effects of such things as friction and air resistance.

The first successful theory of mechanics grew out of an understanding of the role played by time in physical processes. As we will see in subsequent chapters, practically all of physics is concerned with time. There are few things in the world that are perfectly static. Physical processes don't exist; they happen. If one is willing to make allowance for a few minor exceptions, it would even be possible to define physics as the science that studies physical changes. By definition, change is something that takes place in time.

CHAPTER 4

Calculus and the Idea of Determinism

GALILEO STRUGGLED WITH THE problem of the nature of motion for decades. Of all the questions that he had to answer, the one that he found most difficult was that of the nature of instantaneous velocity. For years, the concept appeared to him to be nothing more than a mathematical fiction, if not a contradiction in terms. Once he realized that instantaneous velocity was a quantity with which he had to deal if he was to explain the behavior of moving bodies, the concept continued to cause problems.

In order to see why this concept caused Galileo so much mental anguish, we must ask, "How is this quantity we call 'velocity' to be defined?" This sounds like a trivial question. After all, we all have an intuitive idea of velocity. However, if we look at the question in detail, we quickly discover that things are not as simple as they seem.

If a body moves at constant velocity, there are no problems. Velocity is simply distance divided by time. If a ball is rolled along the ground at a constant rate, and if it moves ten feet in two seconds, then its speed is five feet per second.

The determination of average velocity is no problem either. For example, an object that is dropped from a height of sixty-four feet will fall to the ground in two seconds. During those two seconds, its speed constantly changes as the object is accelerated by gravity. Nevertheless, we can again divide distance traveled by time, and say that its average velocity is thirty-two feet per second.

But suppose we ask how fast the object was moving after exactly one-half second had elapsed. In this case, the definition of velocity as distance divided by time no longer seems to work. During an instant in time, the object does not move any distance at all. An "instant," after all, is defined as a time period of zero duration.

To Galileo, the concept of velocity implied motion. Yet instantaneous velocity was velocity during a period in which the object did not move. To ask what instantaneous velocity was seemed to be equivalent to asking how far an object moved during a time that was so short that the motion was frozen.

Galileo never did come up with an adequate definition of instantaneous velocity. He sidestepped the question by imagining a situation in which a body attained a certain instantaneous velocity and then continued moving at the same speed. Instantaneous velocity, in other words, was the velocity that one would measure if it was constant and *not* instantaneous. For example, the velocity attained by a falling object after one-half second was set equal to the velocity that it would have had if it was not accelerated further.

Galileo should not be criticized for making use of such a muddled definition. On the contrary, he deserves praise for attempting to deal with quantities that were changing in time. That he could obtain the correct formula, $v = at$, for the instantaneous velocity of a falling object in a day when mathematical methods capable of dealing with rates of change had not yet been developed is evidence of genius, not cause for censure.

It is easy to deal with constant quantities. For example, the height or width of an object can be measured with a ruler. Weight can be measured by placing an object on a balance or a scale. Nor does the measurement of the passage of time present any great problems. In Galileo's day, clocks that possessed second hands did not yet exist. However, time could be measured with hourglasses, with water clocks, and by feeling the human pulse.

But changing quantities could not be measured directly. Nor was it easy to describe them mathematically. The branch of mathematics that deals with rates of change—calculus—had not yet been invented.

Calculus was developed independently by the English mathematician and physicist Isaac Newton and by the German philosopher Baron Gottfried Wilhelm von Leibniz during the second half of the seventeenth century. This discovery by Newton and Leibniz was one of the greatest mathematical achievements of all time. It opened up entire new fields in mathematics, and in the physical sciences as well. Today, all of physics and most higher mathematics are based on calculus.

Newton and Leibniz had numerous forerunners. Ancient Greek mathematicians, such as Antiphon the Sophist, Eudoxus, and Archimedes, solved some of the simpler problems in calculus. Galileo himself was unknowingly using calculus when he derived his formula, $d = \frac{1}{2}at^2$, for the distance traveled by a falling body. However, it was Newton and Leibniz who worked out the fundamental mathematical theorems, and who made calculus into a discipline that other scientists and mathematicians could use. The Greek mathematicians, Galileo, and other of Newton's and Leibniz's forerunners never understood the implications of the techniques they had stumbled upon. Nor did they gain an entirely adequate understanding of the problems concerning rates of change and instantaneous quantities.

In order to see exactly what it was that Newton and Leibniz did, perhaps it would be best to return to the problem of instantaneous velocity once again. But this time we need not confine ourselves to a consideration of the velocity of a falling body. It might be more illuminating to look at the problem of instantaneous velocity in general.

Suppose that an object is moving, and that its velocity

is changing in time. It doesn't make any difference what the object is. It might be a planet moving in its orbit around the sun. It could be a bullet that has been fired by a gun, and that is being slowed down by air resistance. It could even be a leaf that has been picked up by a gust of wind.

Suppose that we want to know what the velocity of the object is at a time that is exactly one second after some arbitrary starting point. How would we determine this quantity? We could obtain a crude estimate of the instantaneous velocity by measuring the distance the object moves in two seconds. We could determine its position at time zero, and again after two seconds had elapsed. This would give us an average velocity that was roughly equivalent to the instantaneous velocity we wanted. After all, the instant with which we are concerned lies at the midpoint of the two-second time interval.

However, it is obvious that there are ways of improving our estimate. If we use the interval from one-half second to one and one-half seconds, and apply the same method, then we should obtain a result that is a little more accurate. But there is no reason to stop here; we can go even further and use the interval that runs from 0.9 seconds to 1.1 seconds, and get an estimate that is even better. In fact, there seems to be no limit to the precision that we can achieve, provided that our measuring instruments are accurate enough. For example, we can take the time interval that runs from 0.9999 seconds to 1.0001 seconds, and compute an average velocity for that.

The averages that we calculate in this manner get closer and closer to the instantaneous velocity that we want. But even if we begin measuring a billionth of a second before the one-second mark, and stop a billionth of a second after it, we still only have an average velocity. It seems that although we have found a way to estimate instantaneous velocity to any degree of accuracy that we desire, we are still as far away from a definition of the quantity as Galileo was.

But suppose we now go a step further and assume that the object travels an infinitesimal distance in an infinitesimal period of time. Even though such small quantities could presumably not be measured, it seems that we could make use of them for the purposes of creating a definition: Instantaneous velocity is an infinitesimal distance divided by an infinitesimal time.

At first glance, it isn't obvious that we have really gotten anywhere. It isn't clear that it is legitimate to deal with such infinitesimal quantities. In fact, it isn't even obvious whether such quantities really exist. On one hand, they have to be larger than zero. After all, the mathematical expression $0/0$ is meaningless; it can be set equal to anything. On the other hand, infinitesimals would have to be smaller than any number one could think of. If they were not, we would not have succeeded in defining instantaneous velocity; we would have an average velocity over some small period of time. It seems that in attempting to define instantaneous velocity, we have had to make use of an idea that is rather suspect.

Newton and Leibniz felt uneasy about the concept of infinitesimals themselves. In fact, more than a century was to pass before mathematicians found a way to dispense with them, and to put calculus on a firm logical foundation. However, it was immediately obvious that the idea, if somewhat suspect, could be extremely useful. The definition of instantaneous velocity as a ratio of infinitesimals had important mathematical consequences. It pointed the way to a new mathematical technique, which is called *calculus* today.

Instantaneous velocity is defined as a rate of change. It is a measure of the rate at which distance changes with time. This means that if one has a formula for velocity, it ought to be possible to find a formula for distance also. The methods of calculus allow one to do precisely this. Newton and Leibniz were able to show, for example, that Galileo's equation for the instantaneous velocity of a falling body, $v = at$, could be used to derive the distance formula, $d = \frac{1}{2}at^2$. Furthermore, it was possible to work in the other direction, and to obtain an expression for velocity if one had a formula for distance.

An equation, such as $v = at$, that contains a rate of

change is called a *differential equation.* If we know what distance a body has traveled at any instant of time, then this equation tells us how far it will travel in the next moment of time. If velocity is viewed as a ratio of infinitesimals, then the equation tells us that in every infinitesimal period of time, the body will travel an infinitesimal distance farther. Calculus is basically a method of adding these infinitesimal distances up.

Perhaps an analogy will make this idea a little clearer. Suppose that one deposits some money in a time deposit in a bank, and that interest is compounded daily. This means that every day, the money will earn a small amount of interest. If one wants to know how much the money will earn in a year, the amounts earned on each day must be added up (there is a formula for doing this, so one really doesn't have to add a column of 365 figures). Similarly, when calculus is used to solve a differential equation, infinitesimal quantities are added up. It is possible to find formulas for doing this also.

The practical significance of this is enormous. For example, if one knows where the earth is at any instant of time, then it is possible to determine where it will be at the next moment of time, and the next moment, and the next, if one can only find a differential equation to describe its motion. And if one knows where it will be at these instants, calculus can be used to determine its motion for *all* moments of time. In fact, it is possible to determine not only where it will be, but also how fast it will be moving, at any time in the future. Furthermore, it is also possible to work in the other direction, and to determine where it was at any time in the past.

The same methods can be applied to any problem in which there are rates of change. Calculus can be used to describe the motion of a vibrating string; it can be applied to the flow of fluids and of electric currents, and to the behavior of particles in the subatomic world. However, it experienced its first success when Newton made use of it in developing his theory of gravitation.

In 1687, Newton published his most important book. Written in Latin, it was titled *Philosophiae Naturalis Principia Mathematica* ("Mathematical Principles of Natural Philosophy"). In this book (which is commonly referred to as *Principia Mathematica,* and sometimes simply as *Principia*), Newton completed the analysis of motion that had been begun by Galileo, stated his law of gravitation, explained the motion of celestial bodies, and delved into such subjects as ocean tides, the motion of bodies that are slowed by air resistance, and the motions of fluids.

At the beginning of the book, Newton stated his three laws of motion. They are so important that it would not be a bad idea to quote them in their entirety:

> Every body continues in its state of rest or of uniform rectilinear motion unless compelled to change its state by forces impressed upon it.
>
> The change of motion is proportional to the impressed force and takes place in the direction of the straight line along which the force acts.
>
> To every action there is always opposed an equal reaction: or, the mutual actions of two bodies upon each other are always equal, and directed to contrary parts.

The first law, which was known to Galileo, is the law of inertia. It states that a motionless object will remain at rest if no forces act upon it, and that an object moving in a straight line will continue on in that direction in the absence of forces.

The third law, Newton's principle of action and reaction, is perhaps best illustrated by rocket propulsion. A rocket is accelerated because it ejects expanding gases backward at high velocities. The forward acceleration is the reaction. It should be noted that according to the first law, a rocket will continue moving in the same direction when the propulsion stops, for example when the fuel is exhausted. Another common example of action and reaction is the recoil of a rifle. As the bullet is propelled forward, the gun "kicks" backward against the shoulder of the person firing it.

It would be absurd to say that any one of Newton's three laws is more important than the other two; all three are needed for the analysis of motion. However, the second law may be the most profound. It is really a statement, in words, of a differential equation. When Newton uses the words "change of motion," he is speaking of a rate of change in time. The second law is frequently written in mathematical form as

$$F = ma.$$

Force equals mass times acceleration. But acceleration is nothing other than the rate at which velocity changes in time. A body falling in the earth's gravitational field, for example, will increase its velocity at the rate of thirty-two feet per second every second. If it is initially at rest, it will be falling at an instantaneous velocity of thirty-two feet per second after one second, at a rate of sixty-four feet per second after two seconds, and so on.

If one knows the force acting on a body, one can apply the methods of calculus to the differential equation $F = ma$, and determine the velocity at any point in time. Since velocity is a rate of change too, one can go a step further and determine where the body will be located in space at any moment in the past or future. The fact that acceleration is a rate of change of a rate of change makes it possible to carry out the operation twice.

Newton's laws of motion didn't do much to enhance the understanding of the motion of projectiles in the case where air resistance was so small that it could be disregarded. This problem had been solved by Galileo. Consequently the use of calculus and the laws of motion did not provide anything new. However, the methods that Newton developed did make it possible to solve more complicated problems, such as the determination of the orbital motions of celestial bodies.

Today, Newton is known as the discoverer of the inverse square law of gravitational attraction. This law states that all gravitating bodies attract one another with a force that is inversely proportional to the square of the distance. For example, when the distance is doubled, the gravitational attraction is one-fourth as great (because four is the square of two); when the distance is tripled, the force is one-ninth as great, and so on.

In reality, Newton was not the first to speculate that gravitating bodies might attract one another in this manner. When Newton's *Principia* appeared, a number of his contemporaries, such as the physicist Robert Hooke, the astronomer Edmund Halley, and the architect Christopher Wren, had already guessed that the inverse-square law might be valid. Newton's achievement was not the discovery of this law; rather it was his demonstration that the law could be deduced from the laws of planetary motion that had been discovered earlier in the century by the German astronomer Johannes Kepler.

Newton's mathematical demonstration of the validity of the inverse-square law made use of calculus. What Newton did was this: If the inverse-square law was true, then the force on a body (such as the earth, or Mars, for example) could be calculated. If this known force was then "plugged into" the equation **F** = ma, the acceleration of that body could be computed also. Once the acceleration was known, calculus made it possible to determine how the body's position and velocity changed in time; an orbit could be calculated. Naturally it was also possible to work in the other direction. If one began with Kepler's laws (which described orbital motion), the same calculations could be done in reverse, and the inverse-square law derived.

If Newton had not discovered calculus, it is unlikely that he would have obtained these results. In such a case, he would most likely have been remembered as one of a group of English scientists who speculated about inverse-square attraction, but who were unable to demonstrate that their speculations were valid.

Strangely, Newton made little use of calculus in the

Principia. As he admitted later, he used calculus to work out the mathematical proofs that he needed, and then replaced these with arguments based on complicated geometrical diagrams. Newton apparently found calculus invaluable, but nevertheless distrusted a method that was based on so questionable a concept as the infinitesimal.

Newton's own writings on calculus illustrate his uneasiness about the logical foundations of the method. In his first paper on the subject, published in 1669, he commented that his method would be "shortly explained rather than accurately demonstrated." In his second paper on the subject (1671), he explained calculus in a slightly different way than he had in the first. In his third paper, he criticized his previous work, and gave yet another explanation, which, however, was really no more satisfactory than the previous two.

Newton could not decide whether infinitesimals were to be regarded as fixed quantities or as quantities that varied continuously. He spoke of them as increments that were "as small as possible," and yet he couldn't say precisely how small they were. Nor could he settle on a name for them. He referred to them variously as "indivisibles," as "nascent increments," and as "evanescent indivisible quantities."

Similar difficulties were experienced by Newton's rival Leibniz. Leibniz defined infinitesimals as quantities that were "vanishingly small" or "infinitely small." He seems to have realized almost at once that this idea led to difficulties. An infinitely small number had to be greater than zero, and smaller than any fraction that one could name. In an article that was published in 1689, Leibniz spoke of infinitesimals not as real numbers but as fictitious ones. If anything, this attempt at explanation made matters even worse. After all, how could one use fictitious numbers to calculate real quantities? Apparently even Leibniz didn't know. In an article published six years later, he attacked "overprecise" critics, and made the rather lame observation that excessive scrupulousness should not cause one to reject a method that had proved so useful.

When Leibniz explained the concept of the infinitely small to Queen Sophia of Prussia, she apparently did not find the notion to be an especially difficult one. According to the Scottish essayist and historian Thomas Carlyle, the queen replied that she needed no instruction on that subject; the behavior of courtiers had made her thoroughly familiar with the infinitely little. However, mathematicians tended to exhibit a different kind of reaction. For example, the Swiss mathematicians James and John Bernoulli described Leibniz's papers on calculus as providing "an enigma rather than an explanation."

In order to see what the problem was, let us return, for a moment, to the definition of instantaneous velocity. I explained this concept by asserting that in an infinitesimal time, an object would move an infinitesimal distance. Since velocity is distance divided by time, instantaneous velocity is a quotient of two infinitesimals.

However, this definition does not work if we take the infinitesimals to be very small numbers of the ordinary variety. If we take a period of time equal to a thousandth of a second, we can compute only an average velocity over that time. Nor does it help if we use a period equal to a billionth of a second, or a trillionth of a billionth of a second; we still wind up with an average. Infinitesimals have to be vanishingly small.

To make things even worse, one sometimes encounters infinitesimals of a higher order, quantities that are infinitely small compared to the already infinitely small ones of the first order. For example, acceleration is defined as the rate of change of velocity. In other words, it is a rate of change of a rate of change. Infinitesimals must be used to define instantaneous velocity, and used again a second time when one wants to speak of instantaneous acceleration. Nor is this the end of it all. If there exist second-order infinitesimals, then there is no reason why there cannot also be infinitesimals of the third, fourth, and higher orders. Apparently there is no end to degrees of infinite smallness.

One might think that the physicists and mathematicians of the seventeenth century would have been reluctant to make use of calculus until such time as it could be put on a firm logical foundation. Perhaps they were reluctant. However, they did not stop using it. Calculus was too great a discovery to be set aside. It opened up entire new fields of mathematics, and before long, physicists realized that they could not do without it either.

The majority of the laws of physics are expressed in mathematical form as differential equations, for the simple reason that most of the physical quantities with which physics is concerned vary in time. Without calculus, none of these equations could be solved, and the search for natural laws would become a futile endeavor. After all, it would do no good to discover a law if one was unable to work out its implications.

There are certain types of problems in which rates of change in time do not play a role. One might, for example, want to determine the manner in which atmospheric pressure changes with altitude. But the solution of such problems requires the use of calculus too. The decrease of pressure with increasing distance above the surface of the earth is also a rate of change. Consequently the methods of calculus must again be used. Strictly speaking, a differential equation is an equation that involves a rate of change of any kind. It does not necessarily have to be a time rate of change. Most of the equations of physics do describe processes that take place in time. But there are exceptions.

In 1734, seven years after Newton's death, the British philosopher Bishop George Berkeley published a book entitled *The Analyst Or a Discourse Addressed to an Infidel Mathematician. Wherein It Is Examined Whether the Object, Principles, and Inferences of the Modern Analysis Are More Distinctly Conceived, or More Evidently Deduced, Than Religious Mysteries and Points of Faith. "First Cast the Beam Out of Thine Own Eye; and Then Shalt Thou See Clearly to Cast Out the Mote of Thy Brother's Eye."* (Authors seem to have been enamored of long titles in those days.) Berkeley charged that the mathematicians who made use of calculus were proceeding in an illogical manner. Calculus, he said, was incomprehensible. Furthermore, mathematicians were guilty of using reasoning that would not be allowed in theology. Pointing out that infinitesimals were "neither finite quantities, nor quantities infinitely small, nor yet nothing," Berkeley concluded that they must be "the ghosts of departed quantities." Anyone who used such methods, he observed sardonically, "need not . . . be squeamish about any point in Divinity."

Dozens of mathematicians responded to Berkeley's criticisms, but none of them was able to answer the charges adequately. In 1734, mathematicians did not know what an infinitesimal was either. Many of them were honest enough to admit their bafflement. Some years previously, the mathematician Michel Rolle, a contemporary of Newton, had admitted that calculus seemed to be nothing but a collection of ingenious fallacies. Others later concluded that calculus was fundamentally illogical, but that the errors it contained managed somehow to cancel one another out. When the students of the French mathematician Jean Le Rond d'Alembert questioned the methods that he taught them, d'Alembert could only urge them to go on nevertheless, promising that "faith" would eventually come to them. The French author Voltaire, who popularized Newton's theories in France (while leaving the task of translating Newton's difficult *Principia* to his mistress, the Marquise de Châtelet), summed up the problems quite aptly. Calculus, Voltaire said, was "the art of numbering and measuring exactly a Thing whose Existence cannot be conceived."

The logical difficulties associated with calculus did not, however, prevent scientists from using it to draw far-reaching conclusions. In their eyes, the practical successes of the method provided evidence in support of determinism, the doctrine that all events that take place in the natural world are determined by antecedent causes.

Since the time of Descartes, many philosophers and scientists had thought of the natural world as an enormous, complicated machine, which operated according to mathematical principles and laws of nature. When scientists began to realize how important differential equations were to physics, this belief was strengthened. Differential equations, after all, allowed one to predict the future behavior of any system, provided that one knew the positions and velocities of its component particles at any moment of time.

For example, the solution of the differential equation $F = ma$ allows one to compute the orbit of any astronomical object, provided that one knows the forces that act on it. If one knows where it is located in its orbit at any given moment and how fast it is moving at that time, then its future positions and velocities can be computed. For that matter, its past positions and velocities can be calculated also.

If such calculations could be carried out for a planet, an asteroid, or a comet, then it ought to be possible to do similar calculations for any collection of objects, at least in principle. Furthermore, it should make no difference (or at least none was known in the eighteenth century) whether these objects were very large, or microscopic. It seemed that if the methods of calculus were correct, then one ought to be able to predict the future behavior of anything. And if one could do this, it seemed to follow, at least in principle, that the future behavior of all matter was determined in advance.

The best-known statement of this idea is one that was made by the eighteenth-century French astronomer and mathematician Pierre Simon, Marquis de Laplace. According to Laplace,

> We may regard the present state of the universe as the effect of its past and the cause of its future. An intellect which at any given moment knew all the forces that animate matter and the mutual positions of the beings that compose it, if this intellect were vast enough to submit that data to analysis, could condense into a single formula the

> movement of the greatest bodies of the universe and that of the lightest atom: for such an intellect nothing would be uncertain; and the future just like the past would be present before its eyes.

If no such intellect (or "Laplace demon," as it is often called) exists, this does not lessen the force of the argument. It is perfectly permissible to maintain that the future is rigidly determined without actually providing a method for predicting the future. It should be noted, by the way, that if Laplace's argument is valid, then there can be no such thing as human free will. After all, human brains and bodies are made up of matter also. Consequently, their behavior ought to be describable by the same differential equations that apply to the constituent atoms of inanimate matter.

Naturally, if the methods of calculus were not valid, then Laplace's argument had no validity either. This argument depended upon the assumptions that the behavior of all matter could be described by differential equations, and that these differential equations could be solved, at least in principle. If calculus really was a collection of fallacies, if the assumption that knowing what was happening at one moment of time allowed one to predict what would take place at the next moment of time, and the next, and the next, was not valid, then there was no good reason for accepting the argument for a deterministic universe.

Doubts on this score were finally removed in 1821 by the French mathematician Augustin-Louis Cauchy. In that year, Cauchy published a work entitled *Cours d'analyse,* in which he presented a method for putting calculus on a firm logical foundation.

Cauchy accomplished this by finding a method for doing away with the troublesome concept of the infinitesimal. He showed that calculus could be based on the idea of *limit* instead.

As we have seen, the problem with infinitesimals was that they were supposed to be nonzero quantities that were smaller than any finite number. This problem can be avoided by considering sequences of numbers that approach zero, but which never quite reach it. For example, consider the sequence 1, $\frac{1}{2}$, $\frac{1}{4}$, $\frac{1}{8}$, $\frac{1}{16}$, $\frac{1}{32}$ . . . (the dots after $\frac{1}{32}$ mean that the sequence goes on indefinitely). Since each number is one-half as large as its predecessor, we can get as close to zero as we want if we just continue the sequence long enough. In such a case, we say that zero is the sequence's limit. Note that even though the sequence gets arbitrarily close to zero, it contains no "infinitely small" quantities.

An explanation of the manner in which Cauchy used such sequences, and the concept of limit, to derive the fundamental theorems of calculus would be a little too technical for a book like this. However, it is possible to say that Cauchy's method, which was further refined by the German mathematician Karl Weierstrass about fifty years later, finally removed all doubts that calculus could be given a firm logical foundation.

Does this imply that the doctrine of determinism is valid? No, not necessarily. The question of determinism is still being debated. No one has yet come up with a definitive solution to the problem, and perhaps no one ever will. It may be one of those questions that will always remain a philosophical one because it cannot be answered by science. And of course science has not provided an answer; Laplace's argument is not valid.

Before I explain what the flaws in the argument are, I want to observe that there is a sense in which determinism is a theory of time. If one asserts that the present configuration of the universe logically implies all future configurations, then the future is—in some sense—contained in the present. If there is only one possible future, then there is a sense in which future events already "exist." On the other hand, if there are many possible futures, then future events are nothing more than potentialities. In Laplace's theory, there is no real distinction between past and future, because the present implies them both.

However, those who wish to continue to believe in free

will may do so if they wish, for Laplace's argument contains two distinct flaws. The first is the assumption that the superhuman intellect of which he speaks could know the laws of nature exactly. This assumption may not be valid; today we realize that one cannot be sure that there is any such thing as an exact physical law.

During the seventeenth and eighteenth centuries, scientists believed that they were discovering the physical and mathematical principles that had been used by God in constructing the universe. But today it is generally conceded that the "laws of nature" that science discovers are really only approximations.

As science progresses, it does not discover any ultimate truths. It simply finds better approximations. Galileo's laws of motion were only approximate descriptions of the behavior of falling bodies and of projectiles. In order to derive them, Galileo had to assume that the force of gravity was always constant, and that air resistance could be neglected. But gravity is not constant; it becomes weaker as one moves away from the center of the earth. Although the variation in gravity has little effect on an object that falls from a height of ten or a hundred feet, one cannot say that Galileo's equations are accurate to the last decimal. Similarly, air resistance always slows motion to some extent, even if the effect is small.

Newton's theories were not exact either. Einstein's special and general theories of relativity demonstrate that Newtonian mechanics provides good approximations only when velocities are not too large and gravitational fields not too intense. Nor is there any reason to believe that Einstein provided the last word. We have every reason to expect that physicists will eventually discover theories that give even better approximations.

The task of discovering physical laws might very well be a never-ending process. There is reason to believe that it might be analogous to the sequence of numbers 1, ½, ¼, ⅛, 1/16 . . . Though we approach nearer to our goal at each step, we might never come to the end. If Laplace's demon were not able to progress to the end of the sequence either, it could not make predictions that were valid for all time. The farther it attempted to see into the future, the more the inaccuracies would multiply. In such a case, the argument for determinism could be said to be valid in only a very approximate way.

Second, Laplace's argument assumes that the positions and velocities of all the constituent particles of the universe could be determined exactly, at least in principle. But the discovery of quantum mechanics in the early part of the twentieth century showed that this assumption was not valid.

Quantum mechanics is the theory that deals with the behavior of matter on the atomic scale. One of its most fundamental results is *Heisenberg's uncertainty principle,* named after the German physicist Werner Heisenberg, who deduced it in 1927. The uncertainty principle states that the position and momentum (or position and velocity, since momentum is defined to be the product of velocity and mass) of a small particle such as an atom or an electron cannot be simultaneously determined. Indeed, we can measure either quantity to any degree of accuracy that we like. However, the more precisely we determine one, the greater uncertainty there will be in the other. If we could pinpoint an electron in space, then its momentum (and hence its velocity) would be completely unknown.

There are a number of different ways of interpreting this principle. One could assert, for example, that it does nothing more than place a limit on the accuracy of simultaneous measurements. However, the majority of physicists adhere to the *Copenhagen interpretation* of the principle, which was developed by the Danish physicist Niels Bohr and his colleagues at Bohr's institute of theoretical physics in Copenhagen. The Copenhagen interpretation states that it is meaningless to speak of a subatomic particle as simultaneously possessing an exact position and an exact momentum. "Momentum" and "position" are regarded as concepts that were developed to describe objects in the macroscopic world, but have only limited validity in the atomic realm.

Because it must deal with quantities that contain uncertainties, quantum mechanics is unable to describe the behavior of subatomic particles in a deterministic manner; it is capable of dealing only with statistical averages. For example, if an electron is propelled toward a fluorescent screen, it is not possible to calculate exactly where the electron will strike the screen. One can only predict the probability that it will hit the screen in one spot or another. In order to verify the predictions of quantum mechanics, it is necessary to direct very large numbers of electrons toward the screen. Quantum mechanics predicts a certain kind of statistical distribution in this case; such distributions are observed in experiments.

It appears that we must conclude that Laplace's demon would not be able to compute the future position of any atomic particle. The most that it could accomplish would be to calculate statistical averages. Conceivably, this could be interpreted to be a determinism of a sort. However, it would be a much "softer" kind of determinism than what Laplace had in mind.

It seems that one must conclude that Laplace's argument really doesn't prove anything. It should be noted, however, that the above arguments do not demonstrate that the world is basically indeterministic, or that free will exists. All that I have shown is that the Laplace argument is inconclusive.

Furthermore, it might be mentioned that some physicists have attempted to find ways to bring determinism back into quantum mechanics. They have argued that if quantum mechanics were a complete theory, then probabilities and statistical averages could be dispensed with. So far, their attempts to find a more comprehensive theory have not been very successful. In particular, some recent theoretical results seem to indicate that if one wishes to assume that events in the subatomic world are somehow determined in advance, then some rather bizarre conclusions must be accepted.* Currently, it appears that the adherents of the Copenhagen interpretation are in a strong position. But then no one knows what the future will bring.

Physics has provided no definitive answer to the question of whether or not the future is already determined. Consequently this question is still a philosophical one. The Laplace demon argument does not hold water. However, there are no arguments for indeterminism that are completely convincing either. It appears that we must conclude that science has not answered the most fundamental question about the character of future time. No one knows whether there is just one possible future, or many.

* For example, there is the "many-worlds" interpretation of quantum mechanics, which assumes that the universe bifurcates into a pair of parallel universes every time that a quantum "choice" is made. The reader who is interested in this interpretation is referred to Paul Davies' *Other Worlds* (Simon & Schuster, 1980).

CHAPTER 5

The Discovery of the Past and the Idea of Progress

THE PEOPLE WHO LIVED DURing the Renaissance knew nothing of the idea of progress. Possibly this can be ascribed partly to their reverence for antiquity, and partly to the fact that some of the medieval ideas about time had not yet been discarded. The Renaissance was a period during which the rediscovery of the ancient classics and the myth of a golden age combined to create the belief that the achievements of the classical era were not likely to be equaled in modern times. Since, like their predecessors, the men* of the Renaissance believed that the time that remained was quite limited, they found it difficult to form a conception of social or intellectual progress. Although they sometimes interested themselves in the ancient idea of cyclical time, they generally felt little inclination to question the idea that the world and humanity were in their old age.

Thanks to the invention of the mechanical clock and to the work of poets like Petrarch and scientists like Galileo, time was becoming an abstract entity that could be saved,

* My omission of Renaissance women is intentional; in this age, men were still the ones who dominated intellectual intercourse.

wasted, discussed, and used to regulate everyday life. However, no one had yet realized that time was something that might stretch endlessly into the past and into the future. One might say that the kind of time that expresses itself as hours and seconds had been discovered, but that time that goes on for eons had not.

Through the eighteenth century, the Biblical chronology, which implied that the world was less than six thousand years old, was generally accepted. Although it wasn't until 1650 that Archbishop James Ussher set the date of creation at 4004 B.C., the idea that the world was created somewhere around 4000 B.C. was already common in medieval times. During the Renaissance, then, the world was regarded as something that was not very old.

We can get a better idea of the Renaissance view of the past if we look at Biblical chronology in terms of generations rather than years. "4000 B.C." and "six thousand years" are concepts that are too abstract to be very meaningful. Most of us tend to think of six thousand years as simply "a very long time." In the psychological sense, there is not that much difference between six thousand years and six million; both are far longer than a human lifetime.

But consider this: If the world was less than six thousand years old, and if the chronology given in the Old Testament was true, then Moses' mother, Jochebed, could easily have known Jacob. Jacob could have been acquainted with Noah's son Shem. And Shem probably would have known Methuselah. Methuselah was only 243 years old when Adam died. Since Adam is supposed to have been created only five days after time began, the entire period from the Creation to the Exodus was spanned by no more than five or six generations (admittedly, some of them were long ones; Methuselah is supposed to have lived to be 969 years old). And yet this period was, according to medieval and Renaissance thought, far longer than the time that remained. Is it any wonder that the medievals became preoccupied with eternity, or that the people of the Renaissance lacked any notion of progress?

When the Protestant Reformation got under way in the early sixteenth century, belief in Biblical chronology was strengthened, at least in those nations whose inhabitants accepted the new Protestant faiths. The Protestants rejected the authority of the Roman Catholic Church and emphasized a literal interpretation of the Bible in its place. Where Catholic scholars, since the time of St. Augustine in the early fifth century, had been willing to interpret Biblical passages allegorically, the Protestants insisted that the books of the Bible had been dictated by God, and that they were consequently an authoritative historical record. Martin Luther, for example, ridiculed the Copernican hypothesis that the earth revolved around the sun, and appealed to a literal interpretation of scripture to support his arguments. He pointed out that the Bible stated that Joshua had made the sun—not the earth—stand still. Luther also took 4000 B.C. as the date of Creation, and concluded that the world would shortly come to an end. "The world will perish shortly," he said. "The last day is at the door, and I believe the world will not endure a hundred years."

When "new stars" (actually supernova explosions) were seen in the sky in 1572 and 1604, they were widely interpreted to be omens of imminent destruction. Previously, the constellations had been thought to be changeless. When the supernovae were observed, this was taken to mean that the corruption that was omnipresent on earth had spread to the heavens. And when astronomers observed sunspots through that new instrument, the telescope, this was taken to be evidence that the sun was decaying also.

To the Elizabethan poet John Donne, even the mountains were a symptom of decay. Accepting the then-common idea that the earth had originally been a perfect sphere, Donne spoke of mountains and valleys as "warts and pockholes in the face of th' Earth," and admonished, "Thinke so; but yet confesse, in this the world's proportion disfigured is."

Donne's contemporary, Sir Walter Raleigh, concurred. When Raleigh began writing his *History of the World* in

1604 while a prisoner in the Tower of London (he had been sentenced to death for high treason; a reprieve would later delay his execution until 1618), he accepted the Biblical chronology implicitly. Raleigh's own date for the Creation was 4032 B.C. Since, like most of his contemporaries, he believed that the world was destined to last for six thousand years, he concluded that less than four hundred years remained.

In the 1630s, the author and physician Sir Thomas Browne expressed similar sentiments, and explained why it was useless to concern oneself with such ideas as that of progress. " 'Tis too late to be ambitious," Browne said. "The great Mutations of the World are acted, or Time may be too short for our designes."

The modern conception of the great, perhaps boundless, extent of time was introduced into Western thought in a rather roundabout way. Ideas about time began eventually to change, not because Biblical chronology was suddenly rejected, or because philosophers and scientists began to draw new conclusions about the nature of time, but because there was a growing understanding of the nature of physical law. A new conception of time grew up and began to spread in what was almost an insidious manner.

After Galileo's work became known, philosophers and scientists slowly began to realize that the behavior of physical objects could be described in mathematical terms. This led to the idea that there existed laws of nature that had been established by God to regulate His Creation. This concept, incidentally, was quite new; the idea of a "law of nature" would have been incomprehensible to most medieval thinkers. Its nearest equivalent in medieval times was that of natural law. But natural law had nothing to do with the behavior of physical objects; on the contrary, it was a moral law that governed human behavior.

The new idea that nature itself was subject to laws depended upon the implicit assumption that such laws did not change with time. Indeed, the suggestion that they did would have seemed almost blasphemous. If there were laws of nature, their divine origin implied that they must remain fixed for all time.

If the universe was governed by laws that did not change, then it was hard to believe that the world was decaying in the way that many Renaissance and Reformation thinkers believed. God's creation might eventually be destroyed. However, one had a right to expect that things would go on more or less in the same way for the foreseeable future. A lawful universe should not suddenly fall apart.

Although the argument is a rather complicated one, it proceeds by clear, logical steps. It is an argument that would almost certainly have been accepted by the leading thinkers of the seventeenth century, if it had ever been stated. But, strangely, it does not appear that it was ever made explicit. It was an ideal that was unconsciously assumed.

Unconscious assumptions are part of the thought of every age. For example, if the classical Greeks had not unconsciously assumed that the world could be understood by means of pure thought (as opposed to myth and religion), there never would have been any such thing as Greek philosophy. During the Middle Ages, the all-encompassing authority of the church and the validity of divine revelation were rarely questioned; a belief in their validity was one of the unconscious assumptions of the age. Since the time of Galileo, scientists have been making unconscious assumptions that are philosophical in nature, and that probably cannot be proved. They have assumed, first, that the universe is ultimately comprehensible, and second, that experimentation and the use of mathematical inference are the tools most likely to bring this comprehension about. These ideas seem so natural that most of us would think it silly to question them. However, there is really no way that we can be sure that they are true. For all we know, there may be things about the universe that we will never comprehend.

Modern philosophy is said to have begun with Descartes. So perhaps it is not surprising that he should have been the first to expound the new scientific outlook in detail. It was only after Descartes' writings became known that discussions of the "laws of nature" became common. It was Cartesian philosophy that was largely responsible for promulgating the idea that mankind lived in a universe whose workings were governed by law.

Paradoxically, practically all of Descartes' own scientific theories were wrong. Although he made significant contributions to mathematics, his ideas about physics and cosmology generally turned out to be erroneous. For example, he thought that the motion of the bodies in the solar system was akin to the motion of water in a whirlpool. The planets, he said, were carried around by a celestial vortex. Other vortices, he added, could be found around other stars.

In the end, this particular error stimulated the advance of science more than it impeded it. After Descartes, it was realized that there was a possibility that the universe might be very large, perhaps even infinite, in extent. If the stars were bodies like the sun, then planetary systems might be innumerable. Though the theory of vortices was to be disproved by Newton, Descartes advanced the Copernican hypothesis by displacing the earth even farther from the center of the universe. In Descartes' cosmology, neither the earth nor the sun was the center of God's creation.

Descartes was also the originator of the idea of cosmic evolution. Descartes realized that if the universe was governed by fixed laws, then these laws might also determine its evolution in time. He suggested that the universe was originally in a state of primordial chaos that was "as disordered as the poets could ever imagine." According to Descartes, as time passed, the operation of natural laws caused stars to be created. According to his theory, the formation of stars was inevitable. He seems to have reached the conclusion that the laws of nature operated in a deterministic way long before Laplace invented his famous argument.

According to Descartes' theory, the earth was originally a small star. After a time, clouds similar to sunspots were formed on its surface. Eventually, several layers of these clouds came into existence. As they piled up, the vortex surrounding the earth was diminished, and then destroyed. Thus the earth, together with the air that surrounded it, fell toward the sun until it was captured by the sun's vortex. More time passed, and "the mountains, seas, springs and rivers could form themselves naturally, and metals appear in the mines, and plants grow in the countryside." Presumably, a similar process could take place at many other places in the universe. If Descartes' theory was correct, there could be numerous inhabited worlds.

But as soon as Descartes presented his theory, he was careful to repudiate it. He stressed that he only meant to say that the earth *could* have evolved in this manner, adding that "we know perfectly well that they [the earth and the stars] never did arise in this way." Revelation tells us, Descartes admonished, that the world was created all at once by God.

No one really knows whether Descartes believed in his evolutionary theory or not. There are passages in his writings that make it clear that he had no desire to bring ecclesiastical censure down upon himself. He had heard of the manner in which the church had treated Galileo, and he had no desire to become a martyr to science himself. His repudiation of the theory could have been sincere; he might have been giving it only as an example of the manner in which the laws of nature could operate. On the other hand, he might really have believed it, and inserted disclaimers in order to protect himself against charges of heresy.

However, one thing is clear. Descartes did not share the opinion, expressed by many of his contemporaries, that the world would soon come to an end. On the contrary, he was one of the first advocates of the idea of progress. Apparently he felt that humanity could look forward to a future of indefinite length during which there would be invented "an infinity of devices by which we might enjoy, without any effort, the fruits of the earth and all its commodities." That which was known, he went on, "is almost nothing compared to what remains to be known." In Descartes' view, the future promised advances in medicine as well as in science and technology. Eventually, he said, mankind could free itself "of an infinity of illnesses, both of the body and mind, and perhaps even also the decline of age, if we knew enough about their causes and about all the remedies which nature has provided us."

Descartes spoke of cosmic, technological, scientific, and social evolution. The only sort that he failed to discuss was the evolution of living organisms. But this gap was soon filled by Leibniz, who began to speculate about such matters around the end of the seventeenth century. In his *Protogaea* of 1693, Leibniz pointed out that there were numerous forms of life that had existed in earlier periods of geological time and had since become extinct. He concluded, therefore, that it was reasonable to believe that "even the species of animals have many times been transformed." Furthermore, Leibniz pointed out, evolution might have been going on for all eternity. There were two possibilities, he said. "Either there was no beginning, and the moments or states of the world have been increasing in perfection from all eternity, or there was a beginning of the process." Both possibilities, he noted, were consistent with the idea of continual evolutionary advance.

Leibniz was the philosopher who proclaimed that this was the "best of all possible worlds," the doctrine that Voltaire lampooned in his novel *Candide*. Although Voltaire's book is funny and entertaining, he presents Leibniz's position in a manner that is not quite fair. Contrary to what Voltaire implied, Leibniz did not deny that there was evil in the world. Nor did he claim that the world had already reached a state of perfection. In Leibniz's view, "the best of all possible worlds" was one that could evolve, and become more perfect in time. In this respect, his position was not so very different from that of Descartes.

Although Leibniz did write about the possibility of

biological evolution, he seems to have thought that this kind of evolution, if it did occur, was something that happened in the dim past. For the most part, his evolutionary speculations are rather metaphysical in nature. Leibniz's concept of evolution is generally bound up with the idea of a future advance toward perfection.

However, Leibniz did not neglect biological evolution entirely; indeed, he went so far as to suggest that humanity itself might have evolved. According to his philosophical theories, human souls could not be created or destroyed. On the contrary, they had always existed. Hence it seemed reasonable to assume that at one time, these souls might have inhabited the bodies of man's prehuman ancestors. Speaking of these souls, Leibniz observed:

> I should suppose that souls which will some day become human have, like those of the other species, been in the seeds, and in the ancestors, up to Adam, and have consequently existed since the beginning of things, always in a sort of organized body. . . . But it seems proper, for several reasons, that they should have existed then only as sensitive or animal souls.

It was only after the first men had evolved, Leibniz went on, that these souls were granted reason by God.

Although Leibniz and Descartes did not find it difficult to speculate about long, possibly endless, eons of time, most of their contemporaries continued to feel the constraints of Biblical chronology. The idea that the world was less than six thousand years old was widely accepted well into the eighteenth century.

But here and there, cracks began to appear in this picture. A few individuals were observant enough to realize that if the world was as young as it was commonly supposed to be, then there were items of fossil and geological evidence that were hard to explain. For example, in 1663, the British naturalist John Ray studied the fossil traces of a primeval forest, and found himself troubled. "There follows such a train of consequences," he confided to a friend, "as seems to shock the Scripture-History of the novity of the world."

In 1691, Edward Lhwyd, a Welsh acquaintance of Ray's, wrote to him about the fall of a massive stone from a mountain in Wales. Lhwyd noted that there were thousands of such stones in the two valleys adjoining the mountain, but that only two or three of them had fallen in the memory of any living individual. He realized that this implied that the mountains of the earth must have existed for a long time indeed, unless, of course, one could "refer the greatest part of [the fallen stones] to the universal Deluge."

Ray drew a similar conclusion. Considering the geological changes that must have been necessary if mountains were to have been formed in the first place, he observed that "if the mountains were not from the beginning, either the world is a great deal older than is imagined, there being an incredible space of time required to work such changes . . . or, in the primitive times, the creation of the earth suffered far more concussions and mutations in its superficial part than afterwards."

But the geologists of succeeding generations did not pursue the matter. Instead of following up Ray's speculations about the age of the earth, they continued to attempt to reconcile geological knowledge with Biblical chronology for the next two centuries. By the beginning of the nineteenth century, a geological theory that was later to be dubbed *catastrophism* had become orthodox. According to the catastrophists, the geological formations found on the surface of the earth could be attributed to a series of floods and cataclysms that had taken place at various times during the earth's history. The last of these catastrophes was, naturally, the Biblical Deluge.

The theory of catastrophism was developed during the eighteenth century to explain the existence of sedimentary rocks, which had presumably been deposited during successive floodings, as well as the character of the fossil record that was preserved in these rocks. According to the catastrophists, life on earth had been extinguished during each upheaval, and then created anew. This doctrine of "special creations" was a way of explaining discontinuities in the fossil record without invoking the still-suspect idea of biological evolution.

Once scientists realized that fossils were the remains of living organisms, they were confronted with the problem of explaining why there had been forms of life in the past that no longer existed in the present. The theory of special creation provided a plausible answer. At the time, paleontological evidence was much less complete than it is now, and it was not necessary to infer evolutionary progression from the fossil record if one did not want to. One only had to assume that life on earth had been destroyed at intervals, and that the biological forms that had been extinguished had been replaced by new ones.

There was a problem with this theory, however. Since the fossil record was so diverse, and since so many different kinds of organisms appeared at different levels in sedimentary deposits, it was necessary to assume that quite a large number of catastrophes (twenty-seven, according to one account) had taken place. Some scientists found this hypothesis of multiple cataclysms hard to accept.

The problem was not that this idea contradicted the account of Creation in Genesis. By this time, it was agreed that the Biblical "days" of Creation could be taken to be periods of time of indefinite length. As a result, one could simply assume that all of the catastrophes except the last had taken place before the creation of Adam. The problem was that of discovering mechanisms that would bring all the assumed catastrophes about. If one assumed that the earth had been flooded on numerous occasions, then it was necessary to explain where all this water had come from and where it had gone. To believe in the Noachian Deluge was one thing; to postulate twenty-seven worldwide floods was quite another.

Of course, catastrophism was not the only possible explanation of the geological and fossil evidence. Indeed, by the end of the eighteenth century, there was a rival theory. In 1785, the Scottish geologist James Hutton read his *Theory of the Earth* before the Royal Society of Edinburgh. The society published the book in its *Proceedings* in 1788, and an expanded, two-volume edition was brought out in 1795.

Hutton broke with the catastrophists by maintaining that geological changes took place not suddenly and at long intervals but slowly and continuously. Sedimentary rock was still forming, he said, primarily in the oceans. Meanwhile, rock exposed to air and water eroded, producing gravel and soil. Valleys were created by rivers; they had not been formed during periodic deluges. Finally, according to Hutton, internal forces within the earth caused subsidence and uplift. Mountains were created slowly over long periods of time, and just as gradually they wore away.

Hutton is considered to be the founder of modern geology. It is also possible to view him as the man who discovered geological time. It was obvious that if his theory was correct, then long ages would be required to work changes in the earth. The theory of catastrophism had implied that the world was somewhat more than six thousand years old. However, acceptance of that theory did not lead one to the conclusion that its age was so very great. If there had been periodic catastrophes at intervals of a few thousand years, it might be only something like fifty thousand years old. Hutton's theory, on the other hand, required that the earth be millions of years old, at the very least. If the geological forces that had molded the surface of the earth in the past were the same as those that were operating at present, change could not have come about quickly. As Hutton himself put it, his theory implied that there was "no vestige of a beginning, no prospect of an end."

In a way, Hutton's theory represented a return to the outlook of Descartes. Like Descartes, Hutton assumed that evolution—in this case, geological evolution—was caused by the operation of natural laws that did not change in time. The catastrophists, on the other hand, had had to appeal to rare, extraordinary events (and sometimes divine intervention as well) to explain geological formations that were observed in the present.

Theories, such as Hutton's, that were based on the assumption that geological changes in the past were the result of forces that were still operating in the present were later to be classified as *uniformitarian.* But Hutton's contemporaries paid little attention to his uniformitarianism. They failed to understand that this was the most important element in his theory, and they referred to it and the theories of other Scottish geologists as "vulcanist," because such theories invoked the existence of forces created by volcanic action and by the earth's internal heat. The catastrophists, on the other hand, were dubbed "Neptunists" because they emphasized the formation of sediments during periods of flooding. It was as though the geologists of Hutton's day didn't want to think about the prospect of endless time and avoided doing so by avoiding the main issue, and by addressing themselves to the less important of the questions that Hutton had raised.

Decades passed, and then, in 1830, the British geologist Charles Lyell published the first volume of a work entitled *Principles of Geology.* The impact of this book was enormous, and at last the issue of uniformitarianism had to be confronted. Although Lyell contributed no important new ideas to geology, he was a great systematizer. By amassing large amounts of evidence and collecting it together in one place, he forced his contemporaries to take notice.

Lyell, who had traveled extensively in North America and Europe, discussed in detail all the ways in which geological change could take place. Such processes as volcanic upthrust, sedimentation, weathering, and erosion—all of which were slow processes—could account, he said, for the geological formations that were observed. It was not necessary to assume that there had been catastrophes, or that there had been any geological forces at work in the past that were not observed in the present. Lyell traced certain animal forms through successive geological strata and pointed out that the fossil evidence indicated that if there had been catastrophes, they must have been local, not worldwide.

It would be difficult to overemphasize the significance of Lyell's book. Not only did it revolutionize the science of geology, it also provided the stimulus for Darwin's theory of evolution. *Principles of Geology* was one of the books that Darwin took with him on the sea voyage during which he first formulated his ideas concerning natural selection. Without it, there might have been no Darwinian theory. Darwin himself later admitted, "I always feel as if my books came half out of Lyell's brain."

The significance of Lyell's *Principles* was that it established that there had been sufficient time for evolution to take place. After it was published, geologists began to realize that the earth must have existed for millions, or tens of millions, of years at the very least. By the 1860s, they were speaking of time spans of hundreds of millions. And Darwin's theory required hundreds of millions; his principle of natural selection implied that evolutionary change took place very gradually. If the earth was only a few tens of thousands of years old, as the catastrophists had claimed, then no significant amount of evolutionary change could have taken place.

Darwin did not originate the idea of biological evolution. As we have seen, the concept is an old one that can be traced at least as far back as Leibniz. Nor was Leibniz the only predecessor of Darwin. During the eighteenth century, a number of French scientists, including the mathematician Pierre Louis Moreau de Maupertuis, the naturalist Georges Louis Leclerc, Comte de Buffon, and the naturalist Jean Baptiste de Monet, Chevalier de Lamarck, had all speculated about biological evolution. Maupertuis had even anticipated modern theories of genetics by postulating the existence of hereditary particles. According to Maupertuis, these particles could undergo mutation, causing changes in the physical characteristics of plants and animals.

Charles Darwin's grandfather, Erasmus Darwin, was also an evolutionist. He speculated about the subject in such a way that the poet Samuel Taylor Coleridge coined the word "darwinize" to describe the construction of wild theories. But it was apparently only the manner in which Erasmus Darwin speculated, not the idea of evolution itself, that Coleridge considered outlandish. In 1819, forty years before Charles Darwin's *Origin of Species* was published, Coleridge delivered a philosophical lecture in which he mentioned a theory which "has become quite common even among Christian people, that the human race arose from a state of savagery and then gradually from a monkey came up through various stages to be man."

Charles Darwin's achievement was not the suggestion that species evolved, but the construction of a theory that explained how evolution could take place. Darwin's theory was based on the concept of natural selection. According to Darwin, the individuals that made up every species were engaged in a constant struggle for survival. Those that possessed hereditary traits that caused them to be well adapted to their environments would pass these traits along to their offspring, while those that possessed less favorable traits died or failed to reproduce. The more favorable traits were thus selected.

Darwinian evolution is a three-step process. First, there must be variation in the members of a species. If all individuals were exactly alike, there would be no favorable traits to select. Second, the traits that cause individuals to be different from one another must be hereditary. Third, there must be selection, so that some traits are preserved in succeeding generations, while others disappear. If all of these elements are present, species will change slowly over long periods of time. In other words, they will evolve.

The idea of natural selection, however, was not original

with Darwin either. It is an old concept that can be dated at least as far back as the seventeenth century. Of course, seventeenth-century naturalists did not link natural selection and evolution together. In their view, selection was only a kind of pruning device that ensured that healthy specimens would survive while weaker ones died out. It was something, they believed, that did no more than prevent the members of a species from becoming progressively more unfit over a period of several generations.

Darwin was not even the first to bring the ideas of evolution and natural selection together. Twenty-eight years before *Origin of Species* was published, a Scottish botanical writer named Patrick Matthew had suggested the idea of evolution by natural selection in an appendix of his book *On Naval Timber and Arboriculture*. A few years later, the English naturalist Edward Blyth had the same idea. Apparently Blyth did not realize the significance of it, any more than Matthew did, for he buried the hypothesis in a paper on the subject of animal varieties.

Nevertheless, it is Darwin who must be credited as the originator of the theory of evolution by natural selection. Though he was not the first to form the hypothesis, he was the one who amassed the evidence that was needed in order to demonstrate its validity. His role was not unlike that of Newton, who may not have been the first to think of the idea of an inverse-square law of gravitation, but was the only man in seventeenth-century England who could demonstrate the validity of the idea mathematically.

It is not my intention to discuss Darwin's theory in detail, or to chart the progress of the concept of natural selection. I want only to show that the idea of evolution was "in the air" long before Darwin committed his theory to paper, and to investigate the connection between evolutionary ideas and the Victorian preoccupation with the idea of progress.

A number of authors have suggested that even if Darwin had never lived, a theory of evolution was practically inevitable, given the fact that the idea of progress was almost an incontrovertible article of faith to the Victorians. Since the Victorians tended to see evolution as a kind of biological version of progress, it is likely that someone or another would have convinced them of the validity of the hypothesis of evolution by natural selection if Darwin hadn't.

If evolution was the kind of idea that Victorian culture was predisposed to accept, then one would expect that the theory would have been accepted quite rapidly once it was proposed. Indeed, this is exactly what happened. It is true that there was some opposition to Darwin's theory in the years following the publication of *Origin of Species.* But the extent of this opposition has been exaggerated. Darwin's ideas were not only accepted rapidly, they were practically embraced. In 1863, just four years after the publication of Darwin's book, the English geologist Charles Kingsley observed that the scientific world was in a "most curious" state. Darwin, he said, was "conquering everywhere and rushing in like a flood." The following year, Darwin received the Copley Medal of the Royal Society, the highest scientific honor in England. And in 1866, the botanist Joseph Hooker was able to report happily to the British Association that Darwin's theory had become accepted gospel.

The lay public took to Darwin's theory just as enthusiastically. To be sure, there was a little thunder from the pulpit, and a bit of opposition from creationists. However, the majority of the public were more than ready to accept any new idea that had been certified by the scientific community, especially one that fit in so well with the prevailing ethos. And, of course, evolutionary ideas spread quite rapidly. It could hardly have been otherwise. The Victorian period, after all, was an age of intense interest in science. It was an era during which eminent scientists frequently lectured to working-class audiences, one in which so formidable a treatise as Lyell's sixteen-hundred-page *Principles of Geology* could become a best-seller.

In 1869, the church paper *The Guardian* recommended the tenth edition of *Principles of Geology,* in which Lyell declared his adherence to Darwinian theory, to its readers. This led Darwin's friend the naturalist Alfred Russel Wallace to remark that "what with the Tories passing Radical Reform Bills and the Church periodicals advocating Darwinism, the millennium must be at hand." Darwin disagreed. He was afraid that his theory had become orthodox too quickly, and that a reaction against it might soon set in. He needn't have worried. When Darwin published his *Descent of Man* in 1871, he found that his ideas about human ancestry were accepted with hardly a murmur.

Strictly speaking, Darwin's theory was not a theory of progress. When species evolve, they do not really attain "higher" levels; they simply adapt to changing environments. We humans like to think that we represent a sort of evolutionary peak. But there is really no good reason for accepting such an anthropomorphic sentiment. It is even conceivable that humanity could turn out to be nothing more than a momentary evolutionary aberration. Though the genus *Homo* is only about 2 million years old, we already possess the capability of destroying ourselves. And a genus that lasts only 2 million years is not a very successful one in evolutionary terms. If we want to be objective about the matter, it seems we must conclude that it does not seem likely that we will ever approach the success of the cockroach, which evolved about 250 million years ago.

But the Victorians viewed matters differently. In their eyes, the terms "evolution" and "progress" were practically synonymous. So enamored were they with evolutionary ideas that they sought to find examples of evolution in other areas than in the biological world. They didn't have to look far. Not long after Darwin's theory had been accepted, Victorian anthropologists began to speak of the evolution of human customs and society. They even viewed sexual practices in evolutionary terms. According to the accepted doctrine of the day, members of the earliest human societies—and of contemporary primitive societies as well—had engaged in wholesale promiscuity. Over long periods of time, this had evolved into Victorian monogamy, which was naturally the highest state of all.

Meanwhile, the British philosopher Herbert Spencer was busy elevating evolution into a cosmic principle. According to Spencer, laws of evolution (i.e., progress) governed practically every natural process that could be observed in the universe. The surface of the earth had originally been quite smooth, he claimed. Since then, it had evolved in the direction of greater complexity. Spencer's attitude was precisely the opposite of that of the Elizabethans, who had also thought that the earth was originally smooth, and who viewed the growing irregularity as evidence of corruption. The earth's climate had also evolved over long periods of time, Spencer claimed. Since the earth had been created, it had grown progressively more variegated.

According to Spencer, the principle of evolution was much more than a biological law. The idea could be applied to social phenomena also. Human societies, he said, had also evolved in the direction of greater complexity. The same could be said of human language, and of human tools.

And, of course, the universe itself exhibited the principle of evolution. According to Spencer, stars and planets had evolved from gaseous nebulae. Like everything that it contained, the universe itself grew in complexity over long periods of time. Evolution, in Spencer's eyes, was a universal cosmic law.

Few people read Spencer today. Histories of philosophy sometimes fail even to mention his name. Yet in his own day, he was widely admired, and perhaps the most frequently read of all philosophers. The reason for this is obvious. To an even greater extent than Darwin, Spencer was the prophet of the gospel of progress.

Why were the Victorians so obsessed with the idea of progress? A common answer to this question is that they were impressed by the achievements of the industrial revolution. But there must be more to it than that. Technological achievement proceeds at an even more rapid pace today. And yet we have become skeptics who sometimes wonder if perhaps technology might not be advancing too rapidly, if we might be creating more problems with it than we solve.

Perhaps it is not possible to explain, in any simple way, why progress became almost a religious idea during the latter part of the nineteenth century. The growth of the idea was certainly the result of the interplay of a number of complex factors.

However, it is possible to determine how the idea of progress began. The idea was originally an outgrowth of Descartes' conception of laws of nature. After Descartes, the universe began to be thought of as a great machine that operated according to fixed principles. As a result, the world came to be viewed as something that would continue on into the indefinite future. Belief in divine intervention diminished, and miracles began to be viewed with skepticism. Progress—social, intellectual, technological, and biological progress—was possible because there was *time*.

It was during the Victorian age that this vision of nature reached its peak of influence. During this age, science had more prestige than it had had at any time in history, even more than it has today. The Victorians believed that the discoveries of science provided knowledge that was certain, and that the scientific laws that had been discovered were exact. It was commonly thought that the universe—if not the human mind—was ruled by a Laplacian determinism. Only in the twentieth century would relativity and quantum mechanics begin to introduce uncertainties into this picture.

The prevailing conception of time is a fundamental element in the world view of every age. The ancient Greeks thought of time as something that was cyclical. As a result, they could not conceive of the idea of progress. Instead, they promulgated the myth of a degeneration from a golden age. But in the view of many, the golden age lay not only in the past, but also in the future. For example, the poet Hesiod, who was roughly a contemporary of Homer, comments that he lives in the iron age, the worst one of them all. It would be much better, Hesiod adds, if he had been born much earlier—or later.

The concept of linear time was introduced by Judaism and elaborated upon by the early Christian writers. But it was a linear time of short duration, a time that had begun in the not-too-distant past, which would shortly come to an end. Abstract time began only with the invention of the clock, and with Galileo's investigations of falling bodies. Beginning with Descartes, this time was gradually transformed into the limitless time that is part of the modern outlook.

It has been suggested, by a number of writers, that the reason a technological society developed only in the West was that only the West possessed the concept of linear time. The idea seems to be that linear time and the conception of progress are related. Indeed they are. However, that is only part of the story. Preoccupation with the idea of progress became possible only after time began to be perceived as it was by individuals like Hutton, who saw it as something that had "no vestige of a beginning, no prospect of an end."

CHAPTER 6

The Age of the World

DURING THE SEVENTEENTH CENtury, the theories of Descartes and of Newton accustomed philosophers and scientists to the idea that the universe might be infinite. Although Descartes did not speculate about the matter himself, other authors realized that his vortex theory might imply that space might go on forever. If each star was the center of a vortex that could contain a planetary system, then it seemed reasonable to speculate that the number of such systems might be unlimited.

On the other hand, Newton stated explicitly that he believed the universe to be infinite. He set down his reasons for this belief in 1692, in a letter to the English clergyman and scholar Richard Bentley. If the universe was not infinite, Newton explained, then gravity would cause all of its matter to collect in its center. "But if the matter was evenly disposed throughout an infinite space," he went on, "it would never convene into one mass; but some of it would convene into one mass and some into another, so as to make an infinite number of great masses scattered at great distances from one another throughout all that infinite space."

If the universe might be infinitely extended in space, then it was plausible to suspect that time might be infinite also, that it might extend indefinitely into both the past and the future. To be sure, there was no logical connection between the infinity of space and the infinity of time. But then the human mind does not always require logical connections. Once it begins to contemplate one variety of infinity, other infinities immediately become easier to imagine. Thus it was soon realized, by a few, that a universe without spatial limits might also be one that had no beginning and no end.

While some of the Greek philosophers—Aristotle, for example—had speculated about the eternity of the world, most of the scientists of the seventeenth and eighteenth centuries did not. Christian doctrine put a damper on such speculation. As a result, even those who were familiar with classical Greek philosophy, or with the infinite universes of Newton and Descartes, refrained from discussing the possibility of endless time. The vast majority of the scientists of this period were believers, and they generally avoided expressing thoughts that might seem to be heretical.

But when Lyell showed, during the nineteenth century, that the geological theories that had been based on strict adherence to Biblical chronology were untenable, these self-imposed constraints were soon broken. Although Lyell himself did not believe that time was infinite, only that it was inconceivably vast, some of his followers were less cautious. Soon after the publication of *Principles of Geology,* some geologists began to speak of time that stretched into the infinite past. Since Lyell had shown that the causes of geological change in the past were identical with those that were operating in the present, these scientists had no qualms about leaping to the conclusion that the earth must always have existed.

Such speculation did not go over well with the British physicist William Thomson. The world could not possibly have existed for an infinite length of time, Thomson responded; in fact, it was unlikely that it was more than 100 million years old. Furthermore, Thomson said, the very principles upon which the doctrine of uniformitarianism was based were suspect; they were contradicted by one of the fundamental laws of physics.

Thomson, whom Queen Victoria raised to the peerage in 1892 in recognition of his scientific work, is commonly referred to as Lord Kelvin today. Kelvin's contemporaries generally considered him to be the greatest physicist of the age. Thus when Kelvin objected to the tenets of uniformitarianism, the scientific community took notice.

Kelvin had shown signs of brilliance while he was still quite young. He published his first scientific paper before he was seventeen. By the time he became Professor of Natural Philosophy at the University of Glasgow in 1846, at the age of twenty-two, he had twenty-six papers in print. After he was appointed to the professorship, Kelvin continued to turn out scientific work at a prodigious rate. The significance of his contributions was widely recognized, and he was knighted in 1866. On New Year's Day of 1892, he was made Baron Kelvin of Largs. By this time, his scientific publications numbered more than five hundred, and he had received honorary doctorates from ten institutions in five countries.

Kelvin's objections to uniformitarianism were based on the second law of thermodynamics, which he and the German physicist Rudolf Clausius had discovered independently during the early 1850s. The second law, which is one of the most far-reaching principles in physics, can be stated in numerous different ways. Kelvin's own formulation of it is perhaps the simplest. According to Kelvin, any process that converts energy from one form to another will always dissipate some of the energy as heat. No machine, and no process in nature, can operate with 100 percent efficiency.

Naturally, there is also a first law of thermodynamics. The first law is nothing more than a statement of the principle of conservation of energy: Energy can be converted from one form to another, but it can neither be created nor destroyed.* For example, when a match burns, chemical en-

---

* After Einstein published his special theory of relativity in 1905, this principle had to be modified slightly, for Einstein's famous equation $E = mc^2$ implies that mass and energy are interchangeable. Mass is converted into energy in a nuclear reactor, or in a hydrogen bomb explosion, for example.

ergy is converted into light and heat. An electrical generator converts mechanical into electrical energy; an electric motor is a device that converts it back into mechanical energy again. In none of these processes is energy created or destroyed.

The second law of thermodynamics can be illustrated by the following example: The chemical energy contained in gasoline is converted into mechanical energy in an automobile engine. According to the second law, this process cannot be perfectly efficient. In fact, the efficiency is rather low; the greater part of the energy converted by an internal engine heats the engine or is ejected in the exhaust. Even the most efficient diesel engines convert chemical energy into motion with an efficiency that is somewhat less than 35 percent. Furthermore, there is no way that this limitation can be overcome. According to the second law of thermodynamics, a substantial portion of the energy would be dissipated as heat even in a perfectly frictionless engine.

There is a formulation of the two laws of thermodynamics that is rather flippant, but nevertheless quite accurate. The first law can be stated as "You can't get something for nothing," the second as "Furthermore, you don't even break even." In other words, energy can't be created out of nothing, and when it is converted, something is always lost.

Applying the second law to geology, Kelvin argued that geological forces could not operate at a constant rate for an indefinite period of time. As long as they acted, energy would be dissipated as heat. Eventually, the earth would lose its internal energy and geological change would cease. The store of energy within the earth was not infinite, after all. But an infinite supply of energy was precisely what would be required if geological change had gone on for an infinite time.

Kelvin reasoned that if the second law implied that the doctrine of uniformitarianism was invalid, then he ought to be able to use the first law to calculate how long the earth had existed. Since it was apparent that chemical changes within the earth could produce only insignificant quantities of energy, Kelvin thought it apparent that the temperature of the earth's interior must be gradually decreasing. Energy could not be created from nothing; consequently, the earth's heat must be gradually leaking away.

From the assumption that the earth had originally been created as a molten ball, Kelvin deduced that approximately 98 million years had elapsed since the time its crust had cooled and solidified. Admitting that his result might be affected by uncertainties regarding the structure of the interior of the earth, Kelvin cautiously stated that it might actually be as young as 20 million years, or as old as 400 million.

In Kelvin's day, the theory that the earth had originally existed in a molten state was generally accepted. Today, scientists are not so sure about this. Many think that it condensed from a relatively cool cloud of gas and dust that orbited the primeval sun. But even if this or a similar theory had been known during the nineteenth century, Kelvin's results would not have been affected by it. It was the maximum age of the earth that Kelvin was trying to determine. The assumption that the earth was originally relatively cool would have led to an age that was less, not greater. In this case, there would have been less heat to be dissipated; this could have happened in a shorter time.

Kelvin read his first paper on the cooling of the earth in 1862. Throughout the rest of his life, he intermittently returned to the subject, gradually refining and reducing his estimates over a period of decades. Though he had been willing to admit the possibility of an age as great as 400 million years at first, by 1897 he was maintaining that 24 million was the most probable figure.

Kelvin applied similar ideas to the heat of the sun. In 1862, he calculated that the sun could not have existed for more than 500 million years. His estimate of the most probable figure for its age was 100 million, approximately the same as that he had obtained for the earth. However, Kelvin placed less emphasis upon this calculation than he did on the one he performed for the earth. He realized that since knowledge of solar physics was extremely limited, the argument in this case was less convincing.

Kelvin's attack on the geologists was launched in earnest in 1868. Early in that year, he read a paper before the Geological Society of Glasgow entitled "On Geological Time." Six years had passed since Kelvin had read his first paper on the earth's age. By 1868, he had enough confidence in his method to limit the length of geological time to a span no greater than 100 million years.

Although it was the theories of the uniformitarian geologists that Kelvin was attacking, it was the evolutionary biologists who felt the sting of his argument most keenly. It had long been obvious that if evolution proceeded by natural selection, very long periods of time were required. One of Darwin's most fundamental assumptions had been the postulate that there were no sudden evolutionary leaps. Species evolved, he said, because selection operated upon small variations that came about naturally; evolutionary change took place at a very gradual rate.

When Darwin wrote *Origin of Species,* he gave no numerical estimates of the time required for evolutionary change to take place. Nevertheless, it was obvious to his contemporaries that he was thinking in terms of hundreds of millions, or perhaps billions, of years. Darwin had made one reference to time spans. In the first edition of *Origin of Species,* he had given a calculation that purported to show that it had taken 300 million years for certain geological changes to take place. However, for the most part, he had been content simply to accept Lyell's conclusion that past time had been inconceivably vast.

When Darwin learned that Kelvin had discovered seemingly incontrovertible arguments that 100 million years was the maximum that could be allowed, he was quite naturally troubled. Darwin made private references to Kelvin as an "odious specter," and attempted to modify his theories to accommodate evolution within the new time limitations.

On some occasions, Darwin's attempts to do this caused him to contradict himself. For example, in the last edition of *Origin of Species,* he attempted to speed up the evolutionary process by assuming that "the world at a very early period was subjected to more rapid and violent changes in its physical conditions than those now occurring; and such changes would have tended to induce changes at a corresponding rate in the organisms which then existed." In other words, evolution was more rapid when the world was young and conditions were chaotic. But this statement directly contradicted other passages, carried over from earlier editions of the book, in which Darwin explicitly stated that the evolution of early life proceeded at a *slower* rate.

In the eyes of some scientists, Kelvin's limitations on the age of the earth presented no great difficulties. If 100 million years was all that physics would allow, then 100 million years was the figure with which geologists and biologists would have to work. This was the view that was taken by Thomas Henry Huxley, the British biologist who had led the brief battle for the acceptance of Darwin's theories. In 1869, Huxley remarked:

> Biology takes her time from geology. The only reason we have for believing in the slow rate of change in living forms is the fact that they persist through a series of deposits which, geology informs us, have taken a long while to make. If the geological clock is wrong, all the naturalist will have to do is to modify his notions of the rapidity of change accordingly.

Kelvin disagreed. Although he did not deny that evolution had taken place, he argued that natural selection was too slow a mechanism to bring it about. For that matter, Kelvin wasn't sure that life had evolved on earth in the first place. It might have had an extraterrestrial origin, he suggested. Primitive organisms, he said, could have traveled to the earth from some previously inhabited world. And, of course, there was yet another obvious possibility, according to Kelvin: divine creation.

By the end of the nineteenth century, Kelvin's figure of 100 million years had become practically orthodox. Not only did the geologists accept it, they also made calculations of their own that seemed to confirm Kelvin's estimate. For example, the British geologist John Phillips estimated that sedimentary rock was formed at the rate of one foot every 1,332 years. Taking the thickness of the sediments on the surface of the earth at 72,000 feet, he obtained a figure of 95,904,000 years, which was remarkably close to Kelvin's 100 million. In 1899 the Irish geologist John Joly calculated the amount of salt that was carried into the sea by rivers every year, and concluded that the oceans were approximately 90 million years old.

Today we know that neither the earth nor the oceans are that young. A modern estimate of the age of the earth is 4.6 billion years. It is harder to determine the age of the oceans, but it is currently believed that they are at least 2 billion years old. However, at the end of the nineteenth century, the idea of a 100-million-year-old earth had become dogma. Geologists would neither accept Kelvin's last, presumably most accurate estimate, of 24 million years nor subject questionable calculations to scrutiny when they produced the "correct" result. And few geologists suspected that within the space of a few years, ideas about the age of the earth would undergo a dramatic change.

In 1896, while experimenting with samples of uranyl potassium sulfate, a chemical compound containing the element uranium, the French physicist Henri Becquerel discovered that uranium emitted a mysterious kind of radiation that could blacken photographic plates. About two years later, the Polish-French physicist Marie Curie began to study a number of different natural substances that exhibited the effect reported by Becquerel. In 1898, she and her husband, Pierre Curie, announced the discovery of two new elements, radium and polonium, and Marie coined the term "radioactivity" to describe the spontaneous emission of energy from these substances. Then, in 1903, Pierre Curie and his assistant, Albert Laborde, found that the production of heat accompanied this emission of energy.

In the same year, the British physicist Ernest Rutherford

began to experiment with radioactive substances. Before long he had established that the amount of heat released was proportional to the number of alpha particles emitted. Rutherford had previously established that radioactive substances emitted three kinds of radiation, which he called *alpha, beta,* and *gamma* after the first three letters of the Greek alphabet. Some years later it would be established that beta particles were electrons, and that alpha particles were identical to nuclei of the element helium (composed of two protons and two neutrons).

Although Rutherford, in 1903, did not yet know what alpha particles were, he realized that the energy that was given off when they were released might account for the heat of the earth's interior. Noting that radioactive matter was present in soil at a concentration of about five parts in 10 billion, he suggested that it might very well be present in the same concentration throughout the earth. If it was, he pointed out, it would release enough heat to maintain the earth at a constant temperature. If this was so, all of Kelvin's calculations—which had been based on the assumption that the earth was cooling off—were wrong. The heat produced by radioactivity could keep the earth's interior hot for an enormously long period of time.

In 1904, Rutherford presented his discoveries to an audience at the Royal Institution. As he was about to begin his lecture, he noticed that Kelvin was in the audience. Naturally this presented a problem. Suspecting that the influential Kelvin would not be pleased to hear his theories contradicted by a young, relatively unknown physicist, Rutherford looked for a way out. Apparently he found one. Some years later, he wrote the following account of the confrontation:

> I came into the room, which was half dark, and presently spotted Lord Kelvin in the audience and realized that I was in for trouble at the last part of the speech dealing with the age of the earth, where my views conflicted with his. To my relief, Kelvin fell fast asleep, but as I came to the im-

> portant point, I saw the old bird sit up, open an eye and cock a baleful glance at me! Then a sudden inspiration came, and I said Lord Kelvin had limited the age of the earth, *provided no new source of heat was discovered.* That prophetic utterance refers to what we are now considering tonight, radium! Behold! the old boy beamed upon me.

Nevertheless, Kelvin was not convinced. Shortly after Rutherford's lecture, he published a note rejecting the idea that radium could perpetually emit heat, and argued that the energy it and other radioactive substances released must have been supplied by some external source. "I venture to suggest that somehow ethereal waves supply the energy to radium while it is giving out heat to the ponderable matter around it," he concluded. Two years later, in 1906, Kelvin publicly debated some of his younger colleagues, and denied that radium could account for the heat of either the earth or the sun. He went so far as to claim that heat could not accompany the emission of alpha particles, as Rutherford had reported. The heat, he said, must be the product of the emission of beta particles instead.

The argument did not go on for long. Kelvin died the following year, and the younger generation of scientists quietly continued to study the various kinds of radioactive materials and the types of radiation that they emitted. Before long, it was generally agreed that alpha emission was accompanied by the release of heat, and that an indefinitely old earth was indeed possible.

Several years before Kelvin died, some of these younger scientists had already realized that it might even be possible to go a step further. Radioactive substances might provide a method for determining the age of the earth. The first step toward the development of an accurate *radioactive dating* method was taken in 1905, when the American chemist Bertram Boltwood noticed that lead was formed when uranium underwent radioactive decay. In 1907, following a suggestion of Rutherford's, with whom he had collaborated in the past, he developed the idea that if one measured the amounts of lead and uranium in a mineral it might be possible to determine how old the mineral sample was. If only uranium had originally been present, the amount of lead that was mixed in with it would provide a measure of the amount of time that had elapsed since the mineral sample had been formed. Boltwood measured samples taken from ten localities on three continents, and obtained results that ranged from 410 million to 2.2 billion years. It appeared that the earth was far older than Kelvin claimed. And since the earth could not be younger than the oldest sample, its age had to be 2.2 billion years, at the very least.

If one takes a sample of uranium, all of it will eventually decay to lead. This process is not a simple one. The decay takes place in fourteen steps. It begins when uranium-238 (the number 238 indicates that there are a total of 238 protons and neutrons in the nucleus of this uranium isotope; a uranium nucleus always has 92 protons, but since the number of neutrons can vary, a number of different varieties of uranium, or uranium isotopes, exist) emits an alpha particle, and decays to thorium-234. The thorium emits a beta particle and becomes protactinium-234. Another beta particle is emitted and uranium is created again. But this time it is a lighter isotope, uranium-234. The next three steps produce thorium-230, radium-226, and radon-222, respectively. After a number of further radioactive decays, lead-206 is formed. Since lead-206 is not radioactive, the *decay series* ends at this point.

Although it is not possible to predict exactly when any radioactive atom will decay, it has been determined that exactly one-half of a sample of uranium-238 will decay to lead in 4.5 billion years. Physicists say that the process has a *half-life* of 4.5 billion years. The concept of half-life, incidentally, can be applied to each individual decay, or to the decay series as a whole. If one knows the half-life of the series, it is possible to determine how old a mineral that contains both uranium and lead is. Since it is reasonable to assume that no lead was present originally (chemical and physical processes would cause the original uranium to crystallize by itself), the ratio of lead to uranium allows one to calculate an age to an accuracy of 2 percent or better.

Today, the uranium-lead decay series is only one of several that are used to measure the ages of minerals. Others include the thorium-lead series (half-life of 13.9 billion years), the potassium-argon series (1.3 billion years), and the rubidium-strontium series (47 billion years). It is thus possible to date rocks by several different methods, and to compare results. The well-known carbon dating method, by the way, is useless for dating rocks. It can only be used to determine the age of organic matter that has absorbed carbon from the atmosphere and that is less than seventy thousand or eighty thousand years old.

Use of these methods, and of a more recently developed one called *fission-track dating* (which is based on measurements of the spontaneous fission of uranium nuclei into large fragments rather than the less violent alpha and beta decays), has established that some volcanic rocks are more than 3.7 billion years old. But this is only a lower limit on the age of the earth—3.7 billion years is only an estimate of the date of formation of the earth's crust. The earth itself existed for some time before a solid crust was formed.

Studies of meteorites, which were presumably formed at the same time as the earth, indicate that an age of 4.6 or 4.7 billion years is more likely. This figure has been confirmed by the dating of rocks brought back from the moon by the astronauts who participated in the Apollo missions. Since the moon is smaller than the earth, it is reasonable to believe that its crust solidified at an earlier date. The lunar rock samples should therefore turn out to be slightly older than those found on the earth. This is exactly what is observed. It has been determined that the Apollo specimens have ages of up to 4.2 billion years.

Radioactive dating methods can be used not only to date

rocks that contain radioactive elements but also to determine the age of the radioactive elements themselves. The method used here is a somewhat more complicated one that depends upon comparing the abundances of radioactive elements that belong to different decay series. If the relative abundance of two such elements is compared to a primordial abundance that is calculated theoretically, the ages of the elements can be determined.

The uncertainties associated with this method are naturally somewhat greater than those associated with the dating of terrestrial rocks. Nevertheless, it can be used to obtain a lower bound on the age of the universe. The method cannot be used to determine the date of the big bang* that took place at the creation of the universe, because heavy, radioactive elements were not created then. On the contrary, they were formed by nuclear reactions that took place within the interiors of massive stars. When some of these stars were torn apart by supernova explosions, these heavy elements were scattered through space. Eventually, some of this material became part of the clouds of dust and gas from which the earth was formed.

Since the radioactive elements were formed sometime after the creation of the universe, all that one can say is that the universe must be somewhat older than the elements themselves. The situation is analogous to that of terrestrial rocks. As I noted previously, the discovery of 3.7-billion-year-old rocks tells us only that the age of the earth must be greater than 3.7 billion years; it does not allow us to pinpoint the date of its formation exactly.

The relative abundances of numerous different pairs of radioactive elements has been measured. A minimum age for the universe has thus been calculated in a number of different ways. When all the evidence from this type of radioactive dating is put together, it is possible to conclude that the ra-

* The big bang theory will be discussed in more detail in Chapters 11 and 12.

dioactive elements are at least 10 or 11 billion years old, and possibly as old as 17 billion years. This figure has been confirmed by theoretical determinations of the ages of the oldest stars. Spectroscopic analysis of the light that these stars emit indicates that their age is probably somewhere between 15 billion and 19 billion years. Finally, the age of our galaxy has been estimated to be about 12 billion years. When all this evidence is gathered together, it begins to appear that one must conclude that the universe is probably at least 15 billion years old.

Now, it so happens that there is another method of estimating the age of the universe. Unfortunately, this method does not always give the same result. In fact, during the late 1970s, a group of astronomers concluded that the universe was no more than 10 billion years old. The publication of this result set off a storm of controversy that bore a remarkable resemblance to the one that erupted in the days of Rutherford and Kelvin. Shortly after the turn of the century, scientists were using radioactive dating methods to obtain ages that were greater than Kelvin and the geologists (who continued to cling to the 100-million-year figure for a while) would allow for the earth. Today, other radioactive dating methods are giving ages for the radioactive elements that are longer than those some astronomers will allow for the universe.

In order to see how this controversy came about, it will be necessary to backtrack a bit, and to make a few brief comments about a discovery that was made in 1929. I think that the ensuing discussion will be illuminating. As we shall see, the question of the age of the universe has become a problem on a number of different occasions. Estimates of this age have contradicted one another more than once, and have been subjected to frequent revision.

In 1929, the American astronomer Edwin Hubble announced his discovery that the universe was expanding. Some years previously, it had been discovered that the light emitted by distant galaxies was *red-shifted.* When astronomers looked at certain characteristic wavelengths of light that were emitted by the stars in these galaxies, they found that these wavelengths were somewhat different when the stars were located in galaxies that were very far away.

It had long been known that shifts in wavelength were associated with velocities of approach or recession. Stars that move toward us as they move in their orbits about the center of our galaxy emit light that is shifted toward the blue end of the spectrum, while the light from those that are moving away from the earth is shifted toward the red.

This effect does not really make stars or galaxies look red. The shift affects only the light's component wavelength, not a star's visual appearance. Although the visible light emitted by the star does become redder, some of the invisible ultraviolet wavelengths are shifted too; they are transformed into a visual blue.

But if red shifts are not noticeable to the eye, their existence becomes obvious when one examines light with scientific instruments. One can determine not only whether a star or galaxy is traveling toward or away from the earth, but also the velocity at which it is moving.

The light from a few galaxies is shifted toward the blue. One of these is the great galaxy in the constellation Andromeda, for example. However, all of the galaxies that exhibit blue shifts are close neighbors of our own Milky Way. They are members of what astronomers call the *local group* of galaxies. Since the galaxies in the local group are gravitationally bound, so that they orbit about one another, one would expect that they would exhibit a kind of behavior that is not characteristic of galaxies in general. Indeed, this is exactly what is observed; when one looks at galaxies that lie beyond the local group, the shifts are always red.

Hubble noted that the light from the faintest galaxies was red-shifted by the greatest amount. Since their low apparent luminosity indicated that they were the farthest away, Hubble concluded that the objects that were most remote from our own galaxy were receding at the greatest velocities. He realized that this could mean only one thing: The universe was expanding. The reason that distant galaxies seemed to be traveling away from the earth was simply that the galaxies were receding from one another.

There is a simple analogy that illustrates why this should be the case. Imagine that a lump of raisin-bread dough is placed in an oven. As the dough rises, the distances between individual raisins will increase. Furthermore, the raisins that are farthest apart will recede from ane another at the fastest rate. If two raisins are nearly touching, the distance between them will hardly change at all. But if they lie at opposite ends of the expanding loaf, they will move apart most rapidly.

Hubble was not content to observe that an expansion was taking place. He also wanted to measure the rate at which the universe was expanding. In order to do this, it was necessary to determine the relationship between velocity of recession and distance. Now, finding velocities of recession was quite a simple matter. Once Hubble measured the red shifts, the velocities could be calculated easily. However, the measurement of distances presented a more difficult problem. In fact, there was no way that the distance of any galaxy could be determined directly.

Astronomers use a triangulation method to find the distances of the nearest stars. The apparent position of a nearby star will shift when it is observed from opposite sides of the earth's orbit. Since the earth is about 93 million miles from the sun, its position in space shifts by a distance of 186 million miles over a six-month period. The orbit of the earth is nearly circular, after all, and the diameter of a circle is just twice its radius.

But this method does not work for more distant stars. Nor can it be used to find the distances of other galaxies, which are even farther away. When the diameter of the earth's orbit is small compared to the distance that one is trying to measure, the shift in apparent position is too small to be measurable, even when one is observing through the most powerful telescopes.

Astronomers use many different methods to estimate the

distances of faraway objects. All of these methods are based on the idea that the farther away a source of light is, the dimmer it will appear to be. For example, a candle that is very near to an observer's eye will appear to be quite bright, while light from a distant searchlight will seem to be comparatively less intense.

If one knows what the intrinsic brightness of an object is and compares this to its apparent brightness, it is possible to calculate the object's distance. One can determine how far away a searchlight is if one knows how much light it is emitting, and the very same principle can be applied to a star or a galaxy. When astronomers want to calculate distances, they measure the apparent brightness of a number of different kinds of objects, including bright stars, glowing clouds of gas, supernovae, and even the galaxies themselves.

When Hubble performed his astronomical observations in the 1920s, he placed special emphasis on stars known as *Cepheids.* Cepheids are variable stars. As they pulsate, their brightness increases and decreases in a regular way. The regularity of these changes makes it possible to determine a Cepheid's *period,* the time that elapses between two successive peaks of brightness.

It had been known since 1912 that there was a relationship between a Cepheid's period and its intrinsic luminosity. The brighter a Cepheid was, the more slowly it pulsated. If one measured the period, then it was possible to calculate the brightness, and consequently the Cepheid's distance.

Hubble was able to locate Cepheids in our neighbor, the Andromeda galaxy. Once he had measured its distance, he went on to galaxies that were farther away. When the galaxies were so far away that it was no longer possible to see Cepheids, Hubble simply switched to other distance indicators. For example, if one knows how far away Andromeda is, one can determine the intrinsic brightness of the very brightest stars in that galaxy. If the brightest stars in a more distant galaxy are similar—and there is no reason why they should not be—these stars can be used to determine the distance of that galaxy also. It is necessary to make use of the Cepheids to calibrate the first step of the distance-measuring process. But once that is done, everything else quickly falls into place.

By 1929, Hubble had plotted graphs that showed the relationship between distance and velocity of recession. This, in turn, allowed him to compute the rate at which the universe was expanding. Now, if the universe was expanding, it seemed to follow that there had to have been a time when it was very compressed. A determination of the expansion rate should allow one to work backward and to estimate the amount of time that had passed since the expansion began. Hubble performed just such a calculation. In 1936, he announced that he had determined that the age of the universe could not be more than 2 billion years.

There was just one problem with this figure. By 1936, scientists had already used radioactive dating methods to establish that certain terrestrial rocks were 3.5 billion years old. If Hubble was right, then it followed that the rocks had to be older than the universe itself. Since such a conclusion was obviously ridiculous, it was apparent that either there was something wrong with radioactive dating methods, or Hubble's measurements contained systematic errors.

The discrepancy was finally resolved in the 1950s, when astronomers discovered that there were two distinct types of Cepheid variable, and that their intrinsic luminosities were not the same. By confusing the two types, Hubble had introduced errors into his measurements of nearby galaxies. Since he had measured distances in a stepwise manner, this error had been carried over into the measurement of the distances of more distant galaxies.

During the same period, it was discovered that yet another of Hubble's distance yardsticks was also inaccurate. When looking at galaxies that were so far away that individual Cepheids could not be made out, Hubble had used bright stars as distance markers. It was discovered that on some occasions, he had taken large clouds of glowing hydrogen gas to be especially luminous stars. Thus another systematic error had been introduced.

Corrections to Hubble's results were made, and the age

of the universe was revised upward to approximately 10 billion years. Then, during the 1960s and '70s, further refinements were made in the distance calibrations, and an age of 13 to 18 billion years was obtained.

The age was expressed as a range, rather than as a definite figure, because there were two different kinds of uncertainty that had to be taken into account. The first was a consequence of the fact that distances still had to be measured in a stepwise manner. The distance of the nearest objects was found first. These distances were used to estimate those of objects that were farther away. As a result, any error that was introduced at one step would be carried over to all the rest. Although methods for determining distance had been made somewhat more accurate, the uncertainties were still greater than those that were typical of other kinds of astronomical measurement.

It was thought that the expansion rate could be determined with an accuracy of about 15 percent. However, the uncertainty in the computed age of the universe was somewhat greater than this, because astronomers could not tell exactly how rapidly the expansion was slowing down.

The expansion of the universe is retarded by gravity. The mutual attractions of the galaxies and of other matter in the universe cause recession velocities to decrease over long periods of time. Thus if one wants to determine how old the universe is, it is necessary to find out how rapidly it is expanding now, and also how rapidly it was expanding in the past. The problem of finding the universe's age is similar to that of finding the time that an object has been falling from an unknown height. One must know not only how fast it is moving now but also what the rate of acceleration is. The only difference between this and the problem of finding the age of the universe is that the expansion of the universe is decelerating rather than accelerating.

If one wants to know the deceleration, it is necessary to estimate the amount of gravitational retardation. In order to determine this, one must know how much gravitating matter there is in the universe. Unfortunately, this is a problem that has not been solved.

The problem is a difficult one, because astronomers can only measure the mass of matter that they can see. If an object emits light, or some other type of radiation, such as radio waves or X rays, it is possible to calculate how much mass is present. Unfortunately, the universe contains a great deal of dark matter that gives off no radiation at all. Many galaxies seem to be surrounded by dark haloes. Not only are astronomers unsure as to how much of this dark matter there is in intergalactic space, they do not even know what it is. A number of suggestions have been made concerning the question of what this dark matter might be. The haloes might be made up of black holes, of bodies that are too small to ignite as stars, or even of clouds of subatomic particles called neutrinos. Nor are those the only possibilities. Until astronomers can determine exactly what this dark material is, it is unlikely that they will be able to accurately estimate the gravitational retardation that it causes.

Attempts have been made to measure the deceleration of the universe directly. It is possible to observe galaxies that are more than 10 billion light-years away. Since a light-year is, by definition, the distance that a ray of light will travel in one year, it follows that the light from these galaxies was emitted more than 10 billion years ago. When astronomers look at distant objects, they are also looking great distances into the past. If this is the case, then, in principle, they ought to be able to make a direct determination of the rate of expansion of the universe at a time 10 billion years before the present.

However, this method does not work very well in practice. The uncertainties are even greater than those that are characteristic of distance measurements. Consequently, estimates of the age of the universe derived from such methods are notoriously uncertain. As a result, many different figures are commonly given for the age of the universe. According to some authors, the universe is 15 billion years old; 15 billion years is approximately in the middle of the 13-to-18-billion-year range. Sometimes it is said that the age is 18 billion years; this is an approximate maximum. And sometimes such figures as 10 billion or 20 billion years are given, on the theory that they at least represent the right order of magnitude.

If there was some uncertainty in a figure of 13 to 18 billion years, the estimate seemed at least to be a reasonable one. It was relatively consistent with estimates of the ages of radioactive elements, of our galaxy, and of old stars. Thus, by the middle of the 1970s, it was generally believed that the problem had been solved. A somewhat rough, but reasonably consistent, age for the universe had been obtained.

But then, during the late 1970s, a group of astronomers from the University of Arizona, the Smithsonian Astrophysical Observatory, and Kitt Peak National Observatory threw a monkey wrench into the calculations. They observed that gravitational attraction was causing the local group of galaxies to fall toward a supercluster of galaxies in the constellation Virgo at a velocity of about five hundred kilometers per second. They concluded that if we didn't go into orbit around the Virgo supercluster, we would collide with it in about 60 billion years.

They pointed out, also, that this implied that our galaxy was located in a pocket of space that was expanding less rapidly than the universe as a whole. If this result was correct, then accepted estimates of the rate of expansion of the universe were wrong. Recalculating the age of the universe, they obtained a figure of 7 to 10 billion years.

Naturally this conclusion elicited a great deal of controversy. Other astronomers countered that the new figure could not possibly be correct because it was inconsistent with results that had been obtained for the ages of the oldest stars and for the radioactive elements. Yet others claimed that previous estimates of the rate of expansion were more accurate than the Arizona–Smithsonian–Kitt Peak group claimed.

The controversy has not been resolved. Additional observations have been made, but not all astronomers have obtained the same result. Some data seem to support the older 13-to-18-billion-year estimate. Other evidence seems to indicate that the 7-to-10-billion-year estimate is more accurate. Astronomical distances, it seems, can be measured in a variety of different ways, and different methods are currently giving different results.

The discrepancies are a result of the fact that there is not any general agreement as to which techniques of distance measurement are most accurate. Some astronomers rely on those that were originally developed by Hubble, on the theory that those methods have been refined over a period of decades. Others think that newer techniques, such as measurement of the rotation rate of galaxies, give results that are more precise.* At present, it seems that there are only two conclusions that one can reasonably draw. First, the rate of expansion, and consequently the age, of the universe is uncertain by a factor of about two. Second, if the matter is eventually to be settled, some ideas that are currently well accepted will have to be discredited. If different methods give different results, they cannot all be accurate.

It is not even possible to maintain, with any degree of certainty, that the corroborating evidence obtained from the determination of the ages of stars and radioactive elements makes the 13-to-18-billion-year age more likely. There are uncertainties here too. Furthermore, the fact that the 13-to-18-billion-year age was generally accepted when most of this work was done may have had some influence on the results. One would do well to remember that when it was thought that the earth was only 100 million years old, geologists had no trouble finding evidence that seemed to support that result. Scientific methods are not always as precise as the lay person imagines them to be, and when a result of a certain magnitude is expected, this sometimes influences the outcome.

---

* The rate at which a galaxy rotates is related to the amount of mass that it contains, which is related, in turn, to its luminosity.

How old is the universe? It is almost certainly at least 7 billion years old, and probably less than 20 billion. At present, those who prefer a figure in the upper end of the range can probably be said to have a slight preponderance of the evidence on their side. However, estimates of the age of the universe have been revised so many times in the past that it is not unlikely that new evidence may soon be discovered that will cause them to be changed again. Perhaps the safest thing to do would be to choose 15 billion years as a reasonable mid-range figure, with the understanding that the true age could easily turn out to be 5 or 6 billion years more or less.

CHAPTER 7

Entropy and the Direction of Time

THE BEST WAY TO APPROACH the subject matter of this chapter might be to indulge in a little fantasy. Let us suppose that an extraterrestrial space vessel is drifting through the solar system, and that it passes close enough to the earth to be approached by a space shuttle. Let us suppose, also, that the alien vessel has suffered some sort of catastrophe, and that its crew has been dead for thousands of years.

Although the personnel on the space shuttle have no way of towing the alien ship to earth, they are able to board it and remove a number of artifacts. Included among these artifacts are a number of disks. When these disks are studied by scientists on earth, it is discovered that they are recordings that are analogous to videotapes. Once their function is understood, the scientists have little trouble devising a method of playing them and examining the images that they contain.

There is only one small problem. It seems that the disks can be played in two different directions, and it is not immediately obvious which is the forward direction and which the reverse. The scientists realize, however, that they only need to establish the forward direction in one case. If they find the right direction for one disk, then they can reasonably assume that all the rest should be played back in the same way.

The first "videotape" they examine shows a planet with two moons. The scientists can see that the planet is rotating on its axis, and that the two moons are revolving around it. However, there is nothing in the sequences of images that tells them whether or not they are playing the disk correctly. To determine that, they would have to know whether the planet rotates from east to west or from west to east. Since they do not know the direction of rotation, they cannot be sure that they are not playing the disk backward.

The next disk shows the planet revolving around a star. Again, this tells the scientists nothing. If the disk is being played backward, the planet will simply appear to follow the same orbit in the opposite direction. Both directions of revolution are perfectly compatible with Newton's law of gravitation, and with Kepler's laws of planetary motion.

So the scientists turn to the third disk. They quickly discover that it contains an animated cartoon that depicts the collision of an alpha particle and a thorium-234 nucleus. When the two combine, uranium-238 is formed. Or is it really a cartoon of the decay of a uranium-238 nucleus into thorium-234 and an alpha particle? There is no way to tell, because every nuclear reaction that is observed in nature can also take place in reverse. Whenever it is observed that a certain nucleus can decay by emitting an alpha particle, one can immediately conclude that the absorption of an alpha particle can also cause that nucleus to be formed.

So the scientists go on to the next disk. This one contains another animated cartoon. It shows large numbers of molecules colliding with one another, and with the walls of the enclosure that contains them. By this time, the scientists are beginning to feel rather frustrated, for there is no way to tell whether this disk is being played correctly either. If time were reversed, the molecules of the gas would simply move in the opposite directions. They would undergo the same kinds of collisions, and their motion would be governed by the same physical laws.

By now, the scientists are beginning to wonder if they will ever sort things out. No disk that they have played so far has allowed them to distinguish between the two possible directions of time. However, there are dozens of disks yet to be viewed, so they doggedly go on to the next one.

This disk shows a number of strange-looking vehicles moving around on a paved surface. Again, the scientists are unable to tell whether they are playing the disk correctly, because they do not know which are the front ends of the vehicles and which the back. They realize that if there were only some way to tell whether the vehicles seemed to be emitting heat, then the second law of thermodynamics would tell them whether they were playing the disk backward or not. After all, as Kelvin pointed out, when energy is converted from one form to another, some of it must be dissipated as heat. Unlike molecular collisions, nuclear reactions, and orbital motion, heat dissipation is a process that is not symmetrical in time. There is a difference between heat dissipation and heat absorption.

Unfortunately, one cannot see the flow of heat in a video disk. So the scientists go on to the next. As soon as the picture flashes on the screen that they have rigged up, they realize that this disk is different from all the others. It shows a hot, glowing piece of metal held in a device that resembles a pair of tongs. As the disk is played, the piece of metal grows progressively dimmer; it is obviously losing heat to its surroundings. The scientists realize that at last they have discovered how to play the disks correctly. They know that they are playing this one in the forward direction, because heat always spontaneously flows from hot objects to cool ones. The reverse never takes place. There is nothing in the picture that might be heating the piece of metal up, and they can in fact see heat being conducted from the piece of metal into the tongs. As the former grows cooler, the tips of the tongs begin to glow slightly, and then grow dark again as they lose heat.

The spontaneous flow of heat always takes place in the same direction. When we drop an ice cube into a glass of water, heat will flow from the water into the ice and cause it to melt. We never see ice spontaneously form in a glass of water, unless it has been put in the freezing compartment of a refrigerator. When a piece of hot iron is plunged into water, the iron is cooled, and some of the water evaporates. No blacksmith ever heated iron by putting it into a water bath.

Of course, heat can be made to flow in the opposite direction. If we are willing to expend energy, it is possible to cause heat to flow from a cool object to a warmer one. Indeed, this is the principle on which a refrigerator works. The refrigerator's motor pumps heat from the cool interior into a room at a higher temperature. However, when there is no expenditure of energy, heat will always flow from a warm object to a cool one; the reverse process is never observed.

The statement that heat flows spontaneously in a particular direction is one of the statements of the second law of thermodynamics. It is equivalent to Kelvin's dictum that energy must be dissipated as heat when energy is converted from one form to another.

The equivalence of these two forms of the second law can be demonstrated mathematically. Perhaps the best way to show this equivalence, without using mathematics, is to note that if either statement were incorrect, then perpetual-motion machines would be possible. A rotating wheel could continue revolving forever if its energy of motion were not dissipated as heat by friction. Similarly, if heat spontaneously flowed into hot objects, this heat could be used to run a steam engine. As long as the heat kept flowing, the engine would never stop. In fact, there is a third statement of the second law, expressing this as follows: *A perpetual-motion machine of the second kind is impossible.*

A perpetual-motion machine of the first kind would be one that violated the first law of thermodynamics, the law of conservation of energy. A perpetual-motion machine of the second kind would be one that was not based on the creation of energy out of nothing, but rather on the idea that the basic principles governing the flow of heat could be violated.

The second law of thermodynamics has a chameleonlike character. It can be expressed in more different ways than any of the other laws of physics. The reason that it takes so many different forms is that it is the most general of all the laws that scientists have discovered. It applies to practically everything.

Newton's law of gravitation applies only to bodies that are large enough to exert measurable gravitational forces. The laws of quantum mechanics apply only to microscopic particles. The laws of electricity and magnetism describe the behavior of objects that have electrical charge or possess measurable magnetic fields. The second law of thermodynamics can be applied to any kind of matter, in any form. It is not even necessary that there be a flow of heat. The second law also describes the behavior of chemical, electrical, and mechanical systems in which the production of heat is negligible. It is a law that governs the conversion and transmission of energy, and all matter has energy in one form or another.

Obviously, if the second law of thermodynamics is to be used in cases where there is no heat flow, it must be stated in a more general form. It is not very meaningful to speak of the dissipation or flow of heat when no heat is flowing, or to speak of perpetual-motion machines when one is attempting to describe a chemical reaction, or the production of electrical energy by a dry cell. However, such a general statement of the second law does indeed exist: *The entropy of an isolated system never decreases.*

An isolated system is one that does not interact with its surroundings. Naturally there are no completely isolated systems in nature. Everything interacts with its environment to some extent. Nevertheless, the concept, like many other abstractions that are used in physics, is extremely useful. If we are able to understand the behavior in ideal cases, we can gain a great deal of understanding about processes that take place in the real world. In fact, treating a real system as an isolated one is often an excellent approximation. In many cases, the energy interchanged with the environment is so small that it can easily be neglected. This technique of neglecting unimportant factors has been used since the time of Galileo.

When Galileo sought to understand the behavior of moving bodies, he did not attempt to bring in the effects of air resistance. He realized that in most cases, its effects would be negligible. He also understood quite well that if it did become necessary to study cases where air resistance was important, his theory could be modified later. Indeed, this is exactly what was done by Newton and the physicists of subsequent generations, who developed equations to describe motion in cases where air resistance was not too small to be disregarded.

Physics has been described as "the science of making approximations." Nature is quite a complicated affair. Since it would be a hopeless task to attempt to understand it in all its complexity, physicists simplify matters by neglecting factors they feel are unimportant. Although this sounds like "fudging," the procedure is really quite an effective one. The practice of making approximations allows physicists to discover general principles that would otherwise remain hidden. Then, once the general principles are found, the quantities that were initially disregarded can be brought back in whenever the scientist is confronted with cases where this becomes necessary.

Although no truly isolated systems exist, the assumption that they do is generally a very good approximation. For example, when an ice cube is dropped into a glass of water, a small amount of heat will flow into the glass from its surroundings. However, this interchange of heat will be much less than that between the water and the ice. If one considers the glass containing the water and the ice cube to be an isolated system, for all practical purposes one is only ignoring extraneous processes in order to focus upon what is most important. As a result, one obtains a better understanding of the physical processes taking place than one would if an attempt were made to take everything into account.

The practice of making approximations in order to seek

general principles has led to numerous successes in physics. One of the most important was the discovery of the concept of entropy by Clausius in 1865. Clausius noted that when heat flowed from a hot body to a cool one, something very important was happening. Obviously the heat continued to flow until both subtances were in equilibrium with one another at the same temperature. But something else was going on too.

In order to see what this something else is, we must take a closer look at the manner in which energy causes physical processes to take place. When we do, we realize that energy alone is not sufficient to make things happen. It is also necessary that there be a difference in levels of energy.

For example, the gravitational energy that is contained in water will turn the wheel of a mill or provide energy for the generation of electrical power when it is passed through a turbine. However, it is necessary that the water be allowed to fall from one level to another. No one ever generated power from water in a lake with no outlet, even though water in such a lake would possess gravitational energy. Such energy is unavailable for use, even when the lake is situated far above sea level. In order to use the energy, it is necessary to allow the water to fall. It is not the geographical elevation of the water that is important, but rather the difference in height through which it falls.

Any substance that is warmer than absolute zero (−273° C, the temperature at which all molecular motion ceases) contains heat energy. However, no one has ever devised a way to make a ship run on the heat energy that is contained within the world's oceans, even though the quantity of this energy is considerable. This heat energy cannot be converted into energy of motion when there are no differences in temperature. The heat energy in the oceans is unusable.

Similarly, a steam engine could not operate if the steam that was produced in its boiler were not at a higher temperature than its surroundings. Nor would household electrical appliances run if there were not a difference in voltage in the two wires that connect them to their wall sockets. For that matter, plants could not grow if there were not a flow of energy from the hot surface of the sun to the relatively cool surface of the earth.

In order to see precisely why these differences in energy content are important, we might do well to consider the case of a hot object and a cold one once again. When they are placed in contact with one another, heat will flow from one object to the other, but the total amount of energy that they contain will remain the same. But although the total amount of energy in the system remains the same as the temperatures of the two objects equalize, something is lost.

This something is the ability to do useful work. While there is still a temperature difference, we could use the energy difference to generate electricity, for example. All that we would have to do to accomplish this would be to attach the two ends of a thermocouple to the objects. An electrical current would immediately be created. This current could be used, in turn, to run a small motor. Naturally, after the hot and cold objects came into equilibrium, this would be impossible.

Since it would be awkward to repeat the phrase "ability to do work" over and over again in the course of the ensuing discussion, perhaps I should replace the expression with something simpler. Perhaps the term "disequilibrium" would be the most accurate choice. After all, it is only when disequilibrium—a difference in energy levels—exists that useful work can be done.

Loss of disequilibrium is a characteristic of every natural process. Differences in energy have a tendency to even out and disappear. Water seeks to find its lowest level. Temperature differences average out. The electrical energy that is stored in a battery will leak away if the battery is left unused for too long a time. The nuclear energy that is stored in uranium will also leak away, and dissipate itself as heat, as the uranium undergoes radioactive decay. This radioactive disequilibrium will disappear even more quickly if the uranium is refined and made to undergo a controlled chain reaction in a nuclear reactor. Even the disequilibrium between the hot sun and the cool earth will eventually vanish as the sun burns itself out over a period of approximately 5 billion years.

The Clausius version of the second law of thermodynamics is nothing more than a way of summing all of this up in a single, succinct statement. Entropy can be defined as the absence of disequilibrium. As disequilibrium, or available energy, disappears, entropy increases. Thus it is possible to say that the entropy of any isolated system has a tendency to increase. If we want to be precise, and include exceptional cases where entropy may remain the same, we have the statement of the second law that was given at the beginning of this discussion: The entropy of an isolated system never decreases.

The concept of entropy makes it possible to see the connections between all the different statements of the second law of thermodynamics. Kelvin's observation that energy tends to be dissipated as heat is equivalent to the statement that entropy tends to increase. As the heat dissipation occurs, the amount of useful energy available becomes less. The observation that heat flows spontaneously from hot objects to cold ones is just another way of noting that when such objects are placed in contact with one another, disequilibrium disappears. An increase in entropy is the result. Finally, a perpetual-motion machine of the second kind is impossible because no machine can make use of the energy in its surroundings when there is no state of disequilibrium.

The concept of entropy is an extraordinarily useful one that makes it possible to make use of the second law of thermodynamics in circumstances where there is no exchange of heat. For example, suppose we pump all of the air out of a metal container. It really doesn't make much difference what kind of container it is. It could be a can of some kind, a bottle, or even a metal box. In order to make the example simpler, I will assume that it is a box.

I will assume, next, that the box has a partition that divides it in half. Next, I will imagine that a gas of some sort is allowed to flow into the box on one side of the partition. Half of the box will then be filled with gas, while the other sides remains in a vacuum.

If all this is done, a state of disequilibrium will have been created. There is energy in the side of the box that contains the gas, and none on the side with the vacuum. It would be possible to make use of some of this energy by attaching two pieces of tubing to the box. If the gas is allowed to flow from one side of the box to the other through the tubing, this flow could be used to run a mechanical device of some kind, or perhaps even to generate a small amount of electricity. The principle is the same as that which allows flowing water to generate electricity in a turbine.

Now suppose that a small hole is punched in the partition in the middle of the box, allowing the gas to flow from one side to the other. If this is done, gas will flow until the pressures are equalized. The disequilibrium will disappear, entropy will increase, and the ability to do work will be lost. The process is analogous to the flow of heat when hot and cold systems are brought into contact.

Entropy is often defined by equating it with disorder. Personally, I think that the idea of disequilibrium makes the concept of entropy more intuitively clear. However, the use of the idea of disorder does have the advantage that entropy does not have to be defined in terms of its opposite (defining entropy in terms of equilibrium wouldn't work, because equilibrium is a final state, while entropy can increase as that state is approached).

Disorder—and entropy—increases when an ice cube is dropped into a glass of water and allowed to melt, for example. The orderly arrangement of water molecules in the ice crystal is replaced by the random, more disorderly arrangement that is characteristic of liquid water. Similarly, a situation in which one side of a box contains a gas and the other a vacuum can be said to be more ordered than one in which the gas is dispersed through the entire box.

The disadvantage of thinking of entropy as disorder is that when "disorder" is used as a technical term, it does not always have quite the same meaning that it has when it is used in ordinary language. For example, when a crystal grows in a liquid, *structure* appears that was not present before the process of crystallization began. As the crystal becomes larger, the entropy of the system does increase. The decrease in entropy in the crystal is more than compensated for by increases in entropy in the liquid. Consequently, there is a sense in which the "disorder" of the system increases even as the orderly crystalline structure grows larger. However, one has to be a physicist to see an increase in disorder in this case. The appearance of structure does not always imply an increase in order, even though "structure" and "order" are equated in our everyday speech. Consequently, the identification of entropy with disorder can be confusing to the nonscientist in a few isolated cases. On the other hand, if one defines entropy to be the opposite of disequilibrium, there is never any ambiguity.

The second law of thermodynamics states that the entropy of any isolated system tends to increase. In other words, when things are left to themselves, they have a tendency to run down. But this does *not* imply that entropy increases in every conceivable situation. In fact, decreases in entropy are quite common. Entropy decreases whenever we make ice cubes in a refrigerator, or manufacture an aerosol can that contains compressed gas, or charge a rechargeable battery. Entropy decreases whenever a plant uses photosynthesis to capture some of the sun's energy, and whenever a human being eats a steak.

Although the second law says that entropy will increase in systems that are isolated from their environments, it says nothing about systems that are subject to outside influences. In the above examples, these outside influences are important indeed. A refrigerator is not an isolated system; it possesses a motor that pumps heat from its interior to the room that surrounds it. When gas is compressed into a can, we are not dealing with an isolated system either; energy must be expended to compress the gas.

Nevertheless, the second law is still applicable in such cases. All we need do is to apply it to the larger system that is made up of the object and its surroundings. It is true that entropy decreases inside a refrigerator. But if we take the system of which the refrigerator is a part as a whole, we will find that the second law of thermodynamics is obeyed. Entropy increases in the system that is made up of the refrigerator, the surrounding room, and the electrical circuits that supply the energy needed to run the motor. In the case of the aerosol can, we will find that entropy will increase in the system made up of the can, the device that compresses the gas, and the power source.

Living organisms provide another excellent example. Much is often made of the fact that living organisms store up disequilibrium (in the form of chemical energy in a plant, for example) as they grow. It is sometimes said that organisms exhibit decreasing, not increasing, entropy, and the term "negentropy" (for "negative entropy") has been coined to describe the processes that take place.

While it is true that entropy tends to decrease in living organisms, at least until they die, it is hard to see any great significance in this fact. Organisms, after all, are not isolated systems. Although no one really knows how to go about calculating the amount of entropy in a mushroom, or in a pod of peas, or in a human being, physicists generally tend to be rather certain that the entropy of the system made up of organisms, their environment, and their ultimate source of energy—the sun—does increase. No one has yet shown that the "negentropy" that is supposedly characteristic of life has any great significance.

Whenever we observe that the entropy of a system is decreasing, we can always conclude that the system must be part of some larger system. If we then consider the larger system as a whole, and find that it is reasonably isolated, we can generally be confident that the second law of thermodynamics will apply. Since an increase in entropy is an increase of a quantity in time, the second law allows us to distinguish between the future and the past. It allows us to define what the British astronomer Sir Arthur Eddington called "the arrow of time." The second law tells us that past and future look different; there will be more entropy in the future, and there was less entropy in the past.

None of the fundamental laws of physics makes this distinction. As the examples given at the beginning of this chapter showed, the laws of mechanics, gravitation, and nuclear physics are perfectly symmetrical with respect to time. They cannot be used to determine whether a video disk is being run forward or backward. As far as these laws are concerned, both directions of time look alike. Nor do any of the basic laws that were not mentioned in the examples, such as the laws of electricity and magnetism or quantum mechanics, distinguish between past and future. If time could somehow run backward, they would still provide perfectly adequate descriptions of the behavior of matter.

It should be noted that the second law of thermodynamics says nothing about the "flow" of time. It says nothing about that moment we call "now" that moves inexorably into the future. The second law says only that the universe has a different appearance in the two different directions. For that matter, there is nothing in physics that can be used to describe this flow. Physics can say nothing about the rate at which time "moves past" us (or is it we who move through a motionless time?). The best that one could do would be to say that time progresses at the rate of one second per second or one hour per hour. And, of course, this is meaningless.

I will have a bit more to say about the subjective "flow" of time in the next chapter. For the moment, however, it might be best to return to the discussion of the second law of thermodynamics. When I said that none of the fundamental laws of physics distinguishes between the two directions of time, my exclusion of the second law was intentional. The law of increasing entropy is only a statistical law; it is not "fundamental" because it cannot describe the behavior of an individual atom or molecule; it deals only with the average behavior of large numbers of them. Entropy is not a concept that can be meaningfully applied to a single particle, or even to a small number of particles.

This can be seen quite clearly if we return to some examples that were discussed previously. Consider, for example, the flow of heat from a warm object to a cool one. Now, heat is nothing but molecular vibration. Temperature is a measure of the average velocity of a large number of molecules. When we say that an object is "hot," we mean that, on the average, its constituent molecules are vibrating very rapidly. When an object is "cold," the average motion of its constituent molecules is very slow. However, energy will be distributed among these molecules in a random manner. Some of the molecules in a hot object will vibrate slowly, and some of the molecules in a cold substance will vibrate rapidly. When heat flows into a cold object, we see nothing more than a change in average behavior.

Speaking of heat in terms of molecular vibrations is like talking about average life expectancy. If we take a group of several thousand people, we can predict quite accurately what their average age at death will be. If this were not possible, life insurance companies would go out of business. However, we cannot predict the year of death of any single individual. He may die in infancy, or he may live a hundred years.

Similarly, when we say that heat flows, we mean that a group of rapidly moving molecules gives up some of its energy to a group of slowly vibrating ones, and that there are enough molecules in each group to make statistical predictions possible. If we had only two molecules, the one that possessed the greater amount of energy could easily take energy from the more slowly moving one if they happened to collide in the right way. In this case it would not be possible to speak of heat flow at all, and the concept of entropy could not meaningfully be applied.

The case of the molecules in the box is even simpler. If the gas is originally confined on one side of the partition, and if the partition is then removed, approximately one-half of the gas molecules will quickly move to the other side of the box, and entropy will increase as this happens. But this "one-half" is a statistical average. We cannot predict what any individual molecule will do.

If the box originally contained only one molecule, or a small number of them, anything might happen. For example, four molecules could move about in such a way that all four will be on one side of the box at one moment, and on the other side at another. In such a case, we could not meaningfully say that entropy had increased. After all, the increase in entropy was a consequence of the fact that the molecules were confined on one side of the box at one moment, and more or less evenly distributed through it at a later one.

University of Texas physicist John Wheeler sums up the statistical character of the second law of thermodynamics as follows: "Ask any molecule what it thinks of the second law of thermodynamics and it will laugh at the question." What Wheeler means is that the behavior of a single molecule can only be described by the basic, time-symmetric laws of physics, such as those of mechanics or quantum mechanics. The behavior of an individual molecule is not constrained by the law of increasing entropy. If we are willing to indulge in a bit of anthropomorphism, we could say that an individual molecule has no way of distinguishing between the two directions of time.

This does not imply that "time" is a meaningless concept on the microscopic level. It is clearly very meaningful indeed. It is possible to speak of the velocity of a molecule, and to compute the distance that it travels in a certain period of time. Subatomic particles that decay into other particles have certain lifetimes.* When atoms emit radiation, they are

* These lifetimes are statistical averages similar to the half-lives of radioactive decay.

emitting energy that has a frequency of a certain number of cycles per second.

However, the only method that we have discussed—so far—for defining time's "arrow" seems to disappear when we enter the microscopic world. Consequently there seems to be nothing to prevent us from speculating about the question of whether individual particles might be able to travel backward in time. Such speculation is made possible by the fact that there are no known laws of physics that make backward-in-time motion impossible.

A theory of this sort was proposed by the American physicist Richard Feynman in 1949. Feynman, who was later to receive the Nobel Prize for his theoretical work on elementary particles, suggested that in certain cases, such time-reversed motion might be observed.

Feynman's theory was proposed as an explanation of the behavior of *antiparticles.* Every subatomic particle that is known to physicists has an antiparticle. Particles and antiparticles annihilate one another whenever they collide, and energy is created in their place. For example, when an electron and its antiparticle, the *positron,* happen to meet, they disappear and a gamma ray is seen. The reaction is an example of the transformation of matter into energy, and is described by Einstein's equation $E = mc^2$.

If matter can be transformed into energy in this way, we have every reason to expect that the reverse process should be able to take place also. And indeed it does. Under certain circumstances, a gamma ray can be transformed into a particle-antiparticle pair. If the gamma ray has the right energy, it can disappear and an electron and a positron can pop into existence.

Feynman pointed out that there was another way of looking at this process. Suppose, he said, a positron was nothing more than an electron that was moving backward in time. If such a backward motion could take place, the annihilation and creation events could be interpreted in a somewhat different manner. The "annihilation" could be described as a sudden reversal in the motion of the electron. Perhaps, Feynman said, the electron and positron that appeared to be annihilating one another were really the same particle.

In order to see just what Feynman meant, let us suppose that an "annihilation" takes place at exactly 3:15 P.M. If the electron reversed its time direction at that point, then nothing would exist after that time, except the gamma ray that the electron emitted when it "kicked" itself backward. The electron moved along the time stream to 3:15, and then it began traveling into the past. Nor would the positron exist after 3:15 either.

On the other hand, the electron would be seen twice before 3:15, once as a forward-moving electron, and once as a backward-moving electron. From our point of view, the two particles would seem to be somewhat different from one another. In particular, if we failed to realize that one of them was moving backward in time, we might be misled into thinking that it had some properties that were unlike those of an electron.

An electron has a negative electrical charge, while a positron is positively charged. However, according to Feynman's theory, the positive charge is illusionary. Negative looks like positive when it is moving in a direction of time that is opposite to our own.

The theory describes the creation of an electron-positron pair in a similar manner. Such a "creation" takes place when a backward-moving electron (i.e., a positron) reverses direction and begins to move forward in time again. If the theory is correct, then an electron could zigzag back and forth in time forever; it could appear in numerous different places at the same time. For example, an electron could travel forward to 3:15, turn around and go back to 3:10, go forward to 3:17, go back to 3:11, and so on. In such a case, it would pass 3:12 many times. Each time it passed 3:12 it would have a different spatial position.

Feynman seems to have gotten the idea for the theory when his physics professor, John Wheeler, suggested the idea of just this sort of zigzag motion during a phone conversation. In his Nobel Prize acceptance speech, Feynman related the conversation as follows:

"Feynman," Wheeler said, "I know why all electrons have the same charge and the same mass."

"Why?" Feynman asked.

"Because," Wheeler replied, "they are all the *same* electron."

When Wheeler made the suggestion, he was probably doing nothing more than indulging in a flight of fancy, not proposing a serious hypothesis. However, Feynman realized that the idea could be made into a serious theory and went on from there. It is Feynman who is given credit for the theory, by the way, even though it was based on Wheeler's idea, because it was the former who worked out the mathematical details and showed that the theory could usefully be applied to problems concerning the behavior of elementary particles.

At this point, one might be tempted to object, "How do we know that such backward-in-time motion is really possible?" The answer is, of course, that we don't. The conventional view, according to which electrons and positrons annihilate one another, and Feynman's backward-time theory are mathematically equivalent. They seem to be nothing more than two alternative ways of looking at the same phenomena. A positron moving forward in time and a backward-moving electron have the same properties. It is impossible to distinguish between the two, and the physicist is free to use whichever description is more convenient.

The seemingly paradoxical notion that time-reversed motion might be possible is a consequence of the fact that there is no arrow of time on the subatomic level.* In cases where this arrow disappears, we are free to imagine any kind of time behavior that we like.

Those who dislike contemplating backward-in-time motion can take solace in the fact that Feynman's theory does

---

* Except for one rare, minor exception, which will be discussed in the next chapter.

seem to have one defect. It fails to explain why, if such zigzag motion is possible, there are so many more electrons than positrons. The atoms that make up ordinary matter are constructed from particles, not antiparticles. Electrons are ubiquitous, but positrons are ordinarly seen only in the laboratory. This is not what one would expect to observe if electrons disguised themselves as positrons half the time. And, of course, if there were only one electron, which bounced incessantly back and forth, the numbers of electrons and positrons would be equal.

Feynman brought up this objection himself in his conversation with Wheeler. "But, Professor," he objected, "there aren't as many positrons as electrons."

"Well," Wheeler answered, "maybe they are hidden in the protons or something."

Wheeler's reply, incidentally, probably should not be taken very seriously. If something like this were really the case, then objects could suddenly disappear. Suppose, for example, that the electrons in my body (and the other particles, which could presumably reverse themselves and become antiparticles too) never got farther forward in time than 5:30 P.M. If that happened, I would cease to exist at that point. Of course, all of the particles might not decide to turn around at exactly the same time. But even if they didn't, there could be some point in time beyond which none of the particles in the universe would ever go. That would seem to imply that the universe could gradually lose mass over a period of time, and eventually disappear.

Since the law of increasing entropy is a statistical law, we might expect that fluctuations should be possible and that, in some cases, the increase of entropy should reverse itself for a short period of time. After all, the laws of probability tell us that very improbable events must happen sooner or later, even though they don't take place very often. In the long run, red and black will turn up an equal number of times in a game of roulette. However, runs of red or of black happen every now and then.

If one observed one of these fluctuations taking place, it

would appear that events were running backward in time. A very large statistical fluctuation could, for example, cause all of the gas in a box to collect on one side. A fluctuation of a different sort could make heat flow from a cold body to a hot one. Such events would contradict none of the fundamental laws of physics; the motion of the individual molecules would be entirely in accord with the laws of quantum mechanics.

Such large fluctuations are not observed. The reason is not that they are impossible, but simply that they are too improbable. For example, if a box contained only a hundred molecules of gas, we would have to wait more than 10^{20} years (10^{20} is the number represented by the numeral 1 followed by twenty zeros) before we would see something like this happen. Since 10^{20} years is approximately 10 billion times longer than the present age of the universe, one would have every reason to be surprised if such an event occurred.

Furthermore, any reasonably sized box will contain not a hundred molecules but a number of the order of 10^{23} or 10^{24} of them. The probability that all of them could collect together on one side is a number that is so small as to be practically meaningless. One is fairly safe in making the assumption that such an event will never occur.

But perhaps the improbability of large fluctuations can be better illustrated by taking an example from everyday life. If I drop a wineglass and see it shatter on the floor, I am observing an increase in entropy. The breaking of the glass is one of the one-way processes that is governed by the thermodynamic arrow of time. Incidentally, this is a case where the definition of entropy that is based on the concept of disorder is more useful than the one that uses the idea of disequilibrium. When the glass breaks, a disordered state replaces an ordered one.

A statistical fluctuation that was large enough could presumably cause the pieces of the glass to come together, and then cause the glass to leap upward into my hand. There is no fundamental reason why this could not happen, if one could wait a long enough time. Statistical fluctuations could cause the molecules of air in the room to move in just the right way so as to push the pieces of broken glass together. Other fluctuations could create brief increases in temperature along the broken edges, and the glass could be welded back together. Finally, another fluctuation could create a rush of air under the now-mended glass and cause it to be blown upward.

According to the laws of statistics, such a chain of events could take place. However, the probability would be small indeed. It would be very difficult to calculate this probability exactly, but it would probably be something of the order of one in $10^{10^{25}}$. Now $10^{10^{25}}$ is the number represented by the numeral 1 followed by 10^{25} zeros. This number is so large that if it were written out in full, it would more than fill all the books that have ever been published. In fact, if the various nations of the world were to go on printing books at the present rate for 10 billion years, there still would not be enough books to contain all the digits of this number. It appears that the laws of statistics tell us, "Yes, miracles are possible. However, they are so unlikely that their probability is practically zero."

All this sounds like fantasy. However, the discussion illustrates an important point. Since large fluctuations are extremely improbable, we can be reasonably sure that whenever we see a low-entropy state, it did not happen by chance. Ice cubes are not created by chance fluctuations; they are made in refrigerators. Wineglasses are not created by chance events; they are manufactured.

If large fluctuations took place, the arrow of time would be lost. We could watch a videotape of a melting ice cube and find that we were unable to say whether it was actually melting that was taking place, or whether we were watching a reversed videotape of an ice cube forming spontaneously. The improbability of large fluctuations guarantees that the arrow of time will not disappear in the world around us, even though it may seem to fade away in small, microscopic systems.

The thermodynamic arrow of time may depend upon

statistical averages. However, it is very real. Though the arrow can disappear on the subatomic level, nevertheless the arrow of time is not an illusion. Furthermore, the arrow of time exists not only in the events that happen around us but also in the universe as a whole. In other words, the direction of time is not a local phenomenon. When astronomers look at the universe, they see low entropy in the past, and they are justified in expecting that there will be higher entropy in the future. It appears that though an electron may not be subject to the arrow of time, the cosmos is.

CHAPTER 8

The Five Arrows of Time

ALTHOUGH THE THERMODYNAMIC arrow of time is the most important one, there are actually five different ways in which the direction of time can be distinguished. One of these arrows of time is provided by the expansion of the universe. Although there may be nothing very fundamental about the time asymmetry of this process, it does distinguish between past and future; matter in the universe was more compressed in the past, and it will be more dispersed at some future time. However, one cannot be sure that the expansion will go on forever. If there is enough mass in the universe, gravitational retardation will eventually cause the expansion to stop. Currently, the prevailing opinion among astronomers is that this event will never take place. The universe does not seem to contain enough matter to cause this to happen. However, there are significant uncertainties that must be taken into account. As I noted previously, no one really knows what the dark extragalactic haloes are made of, for example, or exactly how much mass they contain. At present, the best estimates indicate that they do not contain enough matter to cause the expansion to halt. However, astronomical estimates of this sort are notoriously unreliable.

Thus one must conclude that a halt in the expansion, although unlikely, is possible. If the expansion does stop, at some point tens of billions of years in the future, a phase of contraction would subsequently set in. Since time would presumably continue to flow in the same direction when this happened, it seems that one cannot be sure that the arrow of time provided by the expansion has any fundamental significance.

Furthermore, it is not known whether or not this arrow of time has any connection with the thermodynamic arrow. The dynamics of an expanding universe depend upon the gravitational interactions in the universe. Unfortunately, no one knows how to compute the entropy of a gravitational field. Physicists are not even sure whether the concept of entropy can be applied to gravity, or whether it is something that is characteristic of matter alone. Naturally, some attempts have been made to solve this and related problems. However, what results have been obtained so far are controversial. The second law of thermodynamics can be applied to the matter that the universe contains. But the question of whether or not it describes the behavior of the gravitational fields that this matter creates is an unsolved problem.

If the second arrow of time—the expansion of the universe—is one that may or may not have fundamental significance, the third arrow of time is so unimportant that its very existence is a puzzle. There seems to be no good reason why this particular arrow should be observed at all. As far as physicists have been able to determine, its existence has no consequences that are relevant to other physical processes.

Before I can explain what the third arrow of time is, it will be necessary to digress a bit and to make some observations concerning the behavior of elementary particles. A good way to begin might be to recall that in the previous chapter it was pointed out that the laws of nuclear physics allow reactions to take place in either direction. If a certain kind of nuclear decay is possible, its inverse will be possible too. For example, certain nuclei can decay by emitting alpha particles. Whenever this is the case, the same nuclei can be created if one bombards an appropriate nucleus with alpha particles of the correct energy.

Many nuclei decay by emitting beta particles, or electrons (the electron has two names because beta decay was discovered a number of years before the identity of beta particles and electrons could be established). When beta decay takes place, a second particle, an *antineutrino,* is emitted also. The antineutrino is the antiparticle of the neutrino. Neutrinos and antineutrinos have neither charge nor measurable mass,* and they travel at velocities close to, or equal to, the speed of light.

The reverse process is not observed. However, this has little or nothing to do with the direction of time. There is no reason why a nucleus could not simultaneously absorb an electron and an antineutrino. The only reason that we do not see this happen is that such an event would be a very unlikely one. Before it could occur, three particles—a nucleus, an electron, and an antineutrino—would all have to come together in the right place at the right time. Furthermore, all three would have to possess the right amount of energy.

Most likely, this reverse beta reaction does occur, but so infrequently that we never see it. However, there are reactions that are similar to beta decay that involve only two particles, not three. For example, if a neutrino collides with a neutron, a proton and a beta particle will sometimes appear in their place. This and similar reactions are observed to take place in the inverse direction.

Nuclear reactions, then, do not exhibit an arrow of time. Neither do reactions between elementary particles in general. However, there is one exception. There is a particle

* Until quite recently, it was believed that neutrinos and antineutrinos had no mass at all. Recent (and controversial) experiments indicate that this might not be quite true. However, if neutrinos do have mass, this mass is extremely small. Consequently, experiments designed to measure it are extremely difficult. So far, no conclusive results have been obtained.

called the *neutral K meson,* or *kaon,* that does exhibit a time asymmetry.

The K meson is a subatomic particle that is observed only in the laboratory. It is one of the numerous different particles that are created when physicists cause particles to collide with one another at high energies in particle accelerators. The K meson is not a constituent of ordinary matter, and it plays no role in nuclear decays.

There are three kinds of K meson. One is positively charged, one is negatively charged, and one is electrically neutral. Of the three, only the neutral K meson participates in reactions that are not symmetrical in time. It is the only one of the thousands of elementary particles that have been discovered that has been observed to have this property.

Like most elementary particles,* the neutral K meson is unstable. It will decay in less than a millionth of a second after it has been created. This decay can take place in a variety of different ways. A common decay is one in which the K meson is transformed into three *pi mesons,* or *pions.* In another decay scheme, it decays into a pion, a positron, and a neutrino. Theoretical calculations indicate that if K meson decays are to be symmetrical in time—if the laws of particle physics are to be unable to distinguish between one of these decays and an inverse process—then the kaon must always decay into three particles. A decay into two would be one that was not time-reversible.

In 1964, it was observed that a decay into two particles does sometimes take place. The anomalous decay doesn't happen very often. K meson decay results in three particles more than 99 percent of the time. However, the rare two-particle decay does happen, and it can be used to define an arrow of time.

---

* The term "elementary particle" is a misnomer. According to currently accepted theory, most of these particles are actually combinations of smaller, even more fundamental particles called *quarks.* An explanation of the role of quarks can be found in my book *Dismantling the Universe.*

Naturally there have been numerous theoretical attempts to explain why K meson decay should exhibit a time asymmetry. So far, none has been successful. The fact that the anomaly has no relevance to other time-asymmetric processes only makes its existence more mysterious. Since K mesons are not a normal constituent of matter, it is hard to imagine that the arrow of time that they exhibit could have anything to do with the expansion of the universe, or with the subjective "flow" of time, for that matter. Nor is this characteristic related to the thermodynamic arrow of time. Theoretical calculations demonstrate that it is not possible to use K meson decay to bring about a decrease of entropy in an isolated system. Even if this were possible, the connection between K meson decay and entropy would remain a tenuous one.

Although attempts to find connections between the other arrows of time have often led to bafflement, we at least have an idea why these arrows exist; without them, the universe would be a paradoxical place indeed. If subjective time sometimes went backward, or if thermodynamic fluctuations could cause entropy to decrease, reality would have a very strange appearance, to say the least. However, the time asymmetry of the K meson seems perfectly irrelevant. No one has been able to come up with a valid reason why it should be observed. No one knows why the K meson should be able to distinguish between the two directions of time when other particles cannot, or why it makes this distinction somewhat less than 1 percent of the time.

The fourth arrow of time is the electromagnetic arrow. Electromagnetic waves (this category includes light, X rays, radio waves, and ultraviolet and infrared radiation) travel into the future, not the past. For example, when radar pulses are bounced off the moon, the echo is detected a few seconds after the pulse is emitted, not a few seconds before. When we look up at the sun, we see the sun in the position that it occupied a little more than eight minutes previously; it takes about that long for light to make its way from the sun to the earth. We do not see the sun in the position that it will occupy eight minutes in the future. Nor do we see it in both places at the same time.

At first glance, the existence of an electromagnetic arrow of time does not seem especially mysterious. After all, if the arrow did not exist, it would be possible to send signals into the past, in violation of the principle of causality. For example, it would be possible to bet on the outcome of a football game. Then, if one lost, one could send a message to one's past self, instructing him to bet the other way.

One might think that of all the arrows of time, the electromagnetic arrow would be the one that could be understood in the most straightforward manner. Unfortunately, this is not the case. There is an extremely baffling problem.

The laws of *electrodynamics,* a theory developed by the Scottish physicist James Clerk Maxwell around the middle of the nineteenth century, successfully describe all radiation as a mixture of oscillating electric and magnetic fields. Like all of the basic laws of physics, those of electrodynamics are symmetrical in time; the equations do not distinguish between past and future. When one solves these equations, two solutions are always obtained. One of these solutions corresponds to a wave that radiates into the future, the other to one that travels into the past. And yet the wave traveling ino the future is the only one that is observed.

One can't even be sure that this should be considered a defect in the theory. After all, all of the other theories that describe fundamental physical processes are time-reversible too. In any case, electrodynamics is an extremely well-confirmed theory. It successfully describes not only the emission of radiation but all known electrical and magnetic phenomena as well. If one could modify the theory so that it did not seem to predict the existence of radiation traveling into the past (and there is no obvious way to do this), it is likely that the modified theory's predictions would turn out to be wrong in other cases.

Thus we are confronted with the problem of explaining

the electromagnetic arrow of time in some other way. The best way to begin might be to imagine that electromagnetic waves could propagate backward in time. If we do that, we ought to be able to get an idea of what such waves would look like, and go on from there.

Since electromagnetic waves cannot be seen, the best way to proceed might be to look at an analogous kind of wave motion. Suppose that one drops a pebble into a perfectly still pond. When the pebble hits the water, ripples will be created. These will expand outward until they reach the edges of the pond; the ripples will take the form of a series of expanding concentric circles.

Now suppose that the ripples were traveling not into the future but backward in time. What would they look like? In order to answer this question, we must first note that the backward-moving ripples would have to be interpreted from our habitual forward-in-time point of view. They would look exactly the same as the forward-moving ripples would in a videotape that was run backward. From our point of view, the ripples would seem to originate at the edges of the pond. They would contract inward to the point where the pebble hit the water. After the pebble did fall into the pond the water would become still.

There is nothing in the laws of wave motion that would make such an event impossible. The mathematical equations that describe forward-in-time wave motion in water allow backward motion also. Thus this situation is perfectly analogous to that encountered in electrodynamics. The laws of nature permit a kind of motion that is not seen.

The reason that this kind of motion is not observed is not that it is impossible, but simply that it is too improbable. If the edges of the pond happened to vibrate in exactly the right way, they might create inwardly moving ripples that would contract to meet the pebble just as it was dropped. If the ripples were of exactly the right magnitude, they could exactly cancel out the effect of the pebble, and the pond could be still after the pebble hit. But of course the chance that such a thing could happen is as small as the probability that a broken wineglass will come together and leap upward into one's hand.

I have described the case of the contracting ripples as though it were an event that took place in the forward-time direction. But wasn't it supposed to be a backward-in-time event? Indeed it was, but this really doesn't make any difference. Ripples moving backward in time would be precisely equivalent to ripples that moved inward to meet the pebble. The two descriptions are nothing more than two different ways of looking at the same thing. It is no more possible to distinguish between them than it is to distinguish between a positron that moves forward in time and an electron that moves backward.

This point cannot be stressed too strongly. We cannot observe motion into the past. From our point of view, the past is already gone. All that we can see is behavior that looks as though it were taking place in a film or a videotape that was being run in the wrong direction. If we do see something like this, it is always possible to interpret it in two different ways, depending upon whether we want to view matters in our habitual forward-in-time way, or adopt the opposite viewpoint.

We are now ready to consider the case of backward-moving electromagnetic waves. For simplicity, I will look at a case that is analogous to that of the ripples in the pond. Suppose that a star or a light bulb is radiating light. The light waves will travel outward in all directions. The crests and troughs of the waves will define a series of expanding concentric circles. The crests and troughs, by the way, are crests and troughs in the oscillating electric and magnetic fields of which the light is constituted.

Light waves that moved backward in time would appear to be contracting spheres, the analogy of the contracting ripples in a pond. If such backward-moving light waves existed, we could interpret them as waves that originated in the walls of a room, or—in the case of the star—somewhere in outer space, which were absorbed by the light bulb or the star after they had traveled inward.

If only backward-moving waves existed, a star or a light bulb would not appear to be bright. It would not emit any energy; on the contrary, it would absorb energy. Now, it is obvious that stars and light bulbs are not dark, energy-absorbing objects. But the fact that they are not is something that must be explained. If the laws of electrodynamics are valid, such strange behavior should be possible.

There are two possible explanations. One of them is analogous to the explanation that was given for the absence of time-reversed waves in ponds. According to this view, inwardly moving radiation of this sort is possible, but simply too improbable to be observed. One could say that this is the same kind of improbable behavior that is forbidden by the second law of thermodynamics.

For example, if a light bulb were to absorb energy in this manner, the atoms in the walls of the room would have to spontaneously emit light in precisely the correct way so as to produce an inwardly moving wave of just the right type. The walls would not only have to emit light, they would have to emit it in a coordinated way so that the resulting crests and troughs would have a spherical pattern. If one section of the wall produced wave crests that were not perfectly aligned with the wave crests produced by other sections, the spherical pattern would be disrupted. Such a perfect wave pattern could not be produced even in a laboratory equipped with the most advanced electronic and optical equipment.

If we could conclude that time-reversed radiation is not observed because the conditions needed to produce it are too improbable, the electromagnetic arrow of time would not be puzzling. One could simply assume either that it was related to the thermodynamic arrow of time, or that, like the thermodynamic arrow, it was a reflection of the fact that certain events are too improbable to be observed. Unfortunately, matters are not so simple, because there exists another possible explanation of the origin of the electromagnetic arrow.

This explanation has nothing to do, as far as anyone can tell, with thermodynamics or with probabilities.

In 1945, Wheeler and Feynman developed a theory based on the assumption that if time-reversed radiation is permitted by the laws of electrodynamics, then it must exist in nature. According to their theory, the reason we do not see this radiation is that emission and absorption processes cause it to be canceled out.

The original purpose of the Wheeler-Feynman theory was not to explain the absence of time-reversed radiation, but to find a way to overcome certain mathematical difficulties in theories that described the interactions between charged particles and electromagnetic fields. Although Wheeler and Feynman achieved only partial success, their theory aroused great interest among cosmologists, because it suggested a possible connection between the electromagnetic arrow of time and the expansion of the universe.

Wheeler and Feynman began by assuming that electromagnetic radiation could be emitted in both time directions. They then sought to see what the consequences of this assumption would be. Thus they considered the case where the radiation emitted in the ordinary, forward-time direction was eventually absorbed by matter of one kind or another. They found that if this happened, the particles that made up this matter would re-emit radiation. This radiation would also be emitted in both directions. Half of it would continue on into the future, but half of it would begin to travel back into the past.

As a result, there would be two time-reversed waves, one that propagated from the original emitter, and one that rebounded backward from the "absorber in the future." Wheeler and Feynman found that these two waves would exactly cancel one another out. When the wave from the absorber in the future approached the present, it would converge upon the original emitter and *interfere destructively* with the waves that were going into the past.

Destructive interference is a phenomenon that has long

been familiar in optics. When light waves are brought together so that the crests of one wave coincide with the troughs of the other, they cancel one another, and darkness results. Destructive interference is a phenomenon that anyone can observe with no scientific aparatus whatsoever. One need only look at a light source through a space between two fingers. If the fingers are then brought together until they are almost touching, light and dark fringes will be seen. The dark fringes are the result of destructive interference, the light ones of constructive interference, which takes place where crests align with crests.

At this point, it might be a good idea to recapitulate a bit. An emitter in the present emits radiation both into the past and into the future. That which goes into the future is eventually absorbed. The absorber in the future sends a wave back into its past. When this converging wave reaches the present, it cancels out the wave going into the past.

In other words, we have explained why there is no radiation going into the past. But what about the wave from the future that converges upon the present? Why do we not see this?

According to the theory, we do see it. But we do not interpret it as a wave from the future. Remember that it is a *converging* wave *from* the future. From our forward-in-time point of view, it will appear to be a *diverging* wave *toward* the future. In other words, it will appear to be ordinary radiation. When the theory is worked out in detail, one obtains the result that this wave from the future will interfere constructively with the wave going to the future from the present; it will make that wave seem more intense.

Does not this theory imply that radiation should then appear brighter than that predicted by the conventional theory? No, not at all. We must remember that, according to the theory, 50 percent of the energy from the emitter was going into the past and 50 percent into the future. Thus, initially, we have a wave into the future that is only half as bright as that which the conventional theory predicts. But when the converging wave comes back from the future, it interferes constructively in one case and destructively in the other. It cancels out the 50 percent going into the past, and adds the missing 50 percent to the wave going into the future. The net result of all this is that we see nothing but an ordinary forward traveling wave.

It all sounds very complicated. Are all these contortions really necessary? Does one really have to invoke the existence of radiation traveling backward in time? If the net result is going to be that the time-reversed radiation cancels out anyway, why bring it in in the first place?

The reader who is inclined to ask such questions does have a point. However, the Wheeler-Feynman theory has a number of points in its favor. It explains the absence of time-reversed radiation without invoking any assumptions that are not already present in the theory of electrodynamics. Electrodynamics implies that time-reversed radiation should be possible. Wheeler and Feynman take the theory at its word. When they do, they find that this kind of radiation is something that one need not worry about because it cannot be observed. The Wheeler-Feynman theory may or may not be true. However, it certainly provides an elegant solution to the problem of the existence of an electrodynamic arrow of time.

Of course, the theory doesn't work if there is no future absorber. If the cancellation of time-reversed waves is to take place, there must be material in the future that can absorb all of the forward-in-time radiation. This implies, in turn, that the universe cannot expand forever. If it did, some of the radiation that was emitted by stars would never be absorbed; matter in the future universe would become too dispersed.

If, on the other hand, the universe eventually contracts, matter will eventually be in a more compressed state than it is at present. In such a situation, all of the radiation that travels into the future would eventually be intersected by matter that was able to absorb it.

As I noted previously, astronomers and cosmologists currently incline toward the opinion that the universe will continue to expand forever. If they eventually gather evidence that provides a conclusive proof of this supposition, the Wheeler-Feynman theory would be discredited. On the other hand, if they find that the universe contains enough mass to eventually halt the expansion and to cause a phase of contraction to begin billions of years from now, it will not be possible to say that the Wheeler-Feynman theory has been confirmed. It will simply have been shown that the theory is not contradicted by astronomical data.

As the Austrian-British philosopher of science Karl Popper has pointed out, scientific theories can never really be proved to be correct. The most that one can do is to attempt to falsify them by looking for experimental disproofs. The theories in which we have confidence, Popper says, are generally the ones that have survived numerous attempts at falsification.

If we eventually deduce that the universe will enter into a phase of contraction at some point in the future, the Wheeler-Feynman theory will have survived one such test. But in that event, physicists will still be under no compulsion to accept it as true. In such a case, there would still be two possible explanations for the absence of time-reversed waves. In fact, the explanation that appeals to the improbability of time-reversed waves might still appear to be the more reasonable one, for it also explains the absence of time-reversed ripples in water.

On the other hand, the Wheeler-Feynman theory does have one important advantage over its competitor. It suggests that there might be a connection between electrodynamics and cosmology. Thus it is the more all-encompassing of the two possible explanations. It is precisely because questions concerning the possible relationships between the various arrows of time are so baffling that the Wheeler-Feynman theory has a certain appeal. Although it doesn't suggest any connection between the electrodynamic arrow of time and the present expansion of the universe, it does suggest that there is a relationship between that arrow and a future contraction.

The nature of the relationships—if indeed any exist—among the four arrows of time defined by physics is a topic that has been the object of considerable controversy. And it is an area in which there has been a paucity of firm results. However, the relationship between the four arrows of physics and the psychological arrow of time is even more mysterious. It is so mysterious that some philosophers have been led to conclude that time does not really exist.

The problem is not that of finding an explanation for the fact that the arrows of physics and the psychological arrow have the same direction. The fact that we are made of ordinary matter and that our bodies can be considered to be thermodynamic heat engines should be sufficient to explain that. The really baffling question is not "Why do the psychological and physical arrows point in the same direction?" but rather "Why are they so different?"

We are all aware of a subjective "flow" of time. We are conscious of a moment we call "now" that seems to move inexorably toward the future. But physics has no need of the concept of "now." Its laws deal only with the direction of time, and say nothing about a present moment. In other words, there is no "flow" of time in physics; all that physics really tells us is that some videotapes picture impossible events when they are played backward.

Indeed, if we were to attempt to introduce the idea of a "flow" of time into physics, we would immediately encounter problems. Admittedly, it is possible to describe how an object moves in time. An automobile, for example, travels at so many miles or kilometers per *hour,* a bullet at so many feet or meters per *second.* But how fast does a "now" move? That is a question to which physics has no answer.

In physics, time is a dimension. Like the three dimensions of space, it can be represented by a line that extends indefinitely in both directions. But the dimension of time has no privileged instants. Every moment of time exists on an equal footing with every other. For example, if we apply Galileo's laws of motion to the flight of a projectile, we can calculate that it will strike its target so many seconds after it is fired. But it doesn't make any difference whether the projectile is fired at 6:00 in the morning or at 4:00 in the afternoon. The same laws apply whether it is fired on February 14 or September 21. The time of flight is always the same, whether the projectile is fired "now," or whether it was fired at some time in the past, or whether it will be fired at some time in the future.

The difference between the two opposing descriptions of time—the description of time as a moving "now" and the view of time as a dimension—was discussed by the Cambridge philosopher J. Ellis McTaggart* early in this century, long before problems associated with the time attracted a great deal of attention from physicists. In 1908, McTaggart published a paper in which he distinguished between what he called an A-series in time and a B-series. By A-series, McTaggart meant the description of time that depends upon ideas of past, present, and future. By B-series, he meant the distinction between earlier and later.

It is easy to see that the two ways of talking about time are somewhat different. For example, "Elizabeth II is the present queen of England" depends upon the idea of "now," while the statement "The reign of Henry VIII preceded that of Elizabeth I" does not. The first makes reference to a present moment of time. The second is as true today as it was during the first Elizabeth's reign, and it will still be true five hundred years from now.

McTaggart claimed that the two descriptions of time were mutually contradictory, and he concluded that time must therefore be an illusion. Although this conclusion is accepted by few contemporary philosophers, philosophy has not discovered any definitive answers to the questions that

---

* Yes, this is the "profound McTaggart" mentioned in William Butler Yeats' poem, "A Bronze Head."

McTaggart raised. It is not easy to see what McTaggart's B-series (which bears a remarkable resemblance to time in physics) and his A-series (which, since it has a "now," looks very much like subjective time) have to do with one another.

Some philosophers and scientists have tried to solve the problem by claiming that the "now" is only a subjective phenomenon. In their view, there would be no such thing as a present time if there were no conscious beings to perceive it. If conscious perception did not create a "now," they say, then all moments of time would coexist on an equal footing.

In my opinion, this view is not tenable. The "now" does not seem to be purely subjective. In fact, it is easy to imagine ways in which one might attempt to establish its objective reality. If I attend a baseball game, my perception of the ball being hit "now" is not something that exists only in my consciousness. Everyone else in the stadium simultaneously experiences the same perception. Since we can agree that we share the same "now," it appears that we must ascribe a certain kind of objective existence to this present moment. Furthermore, I can photograph an event as it happens. When the film is developed, the picture will confirm that the "now" I experienced and the event that took place at that "now" were objectively real. It appears that the problem of subjective time cannot be done away with as easily as some would like. Most likely, we will have to go on accepting the commonsense view that a "now" exists, even if this creates difficult philosophical problems.

The question of the relationship between the subjective "now" and the time of physics is only one of these problems. Another, which seems to be just as difficult to solve, has to do with the question "Exactly what is this perception of a 'now'?"

It is clear that the subjective present is not a mathematical instant. If it were, we would not be able to hear a clock go "tick-tock." The fact that we hear the "tick" and the "tock" as a unit, not separately, seems to indicate that they are both contained within the psychological present. Similarly, if the subjective "now" had zero duration, we would not be able to hear musical rhythms. Musical beats would make little impression if notes that were separated from one another in time were not somehow simultaneously present in our consciousness. For that matter, if the "now" were an instant, we could not perceive motion. At best, we could only observe that a moving body occupied different positions at different instants of time.

And then there is the problem of the "flow" of time, which is so alien to anything encountered in physics. It is clear enough that this "flow" is related, in some manner, to cyclic biological processes which take place within the body. Experiments have shown that when body temperature is raised by a fever, or by diathermy, the time sense is distorted; time seems to flow at a much slower rate. But then why does this "flow" vary so much when body temperatures remain the same? Why does time pass slowly when we are bored, and quickly when we are occupied? Perhaps this phenomenon has something to do with the rate at which information is processed within the brain. But what? And what does this have to do with the ability of some people to wake themselves up at certain hours? For that matter, how is it that we can often estimate durations precisely even when conditions are such that time does not seem to be "flowing" at its usual rate?

In an experiment performed in 1936, two subjects were shut up in a soundproof room for forty-eight hours and for eighty-six hours, respectively. It was found that at the end of their confinements, they were able to estimate the amount of time that had passed to an accuracy of better than 1 percent. And yet a soundproof room is hardly a very interesting environment. One can only imagine that the subjects found that time passed "slowly" indeed while they were in it.

And yet, this time sense, which proved so accurate in the 1936 experiment, fails dramatically under other circumstances. For example, individuals who have spent periods of weeks or months in underground caves have found, upon emerging, that they had grossly underestimated the lengths of time that they had spent underground. Similarly, it is possible to greatly underestimate or overestimate short durations under certain circumstances. Most of us have had the experience of looking up at a clock after spending some time engaged in an engrossing task, and finding ourselves surprised that so much time had passed.

Research into the nature of subjective time does not seem to be a major preoccupation of contemporary psychology. Anyone who pursues the literature on the subject will encounter numerous references to speculations that were made or to experiments that were performed fifty or even a hundred years ago. As a result, the nature of the subjective arrow of time remains even more mysterious than the questions raised by the existence of four separate arrows of time in physics. So perhaps it would be best to put aside the problem of subjective time for now, and to return to the problem of time as it is viewed by physical science.

CHAPTER 9

Relativistic Time

SUPPOSE THAT A COMET HAS entered the solar system and that its orbit will take it very close to the earth. Suppose, further, that astronomers wish to determine the time at which it passes the orbit of Mars. How would they do this?

At first glance, the problem seems to be a simple one. The astronomers need only observe the comet as it passes the Martian orbit. If they know how far the comet is from the earth when they do this, it is easy to calculate a time. For example, if the comet is 306 million kilometers (about 190 million miles) away, then it will take about seventeen minutes for light to travel from it to the earth. Thus when astronomers see it intersect the Martian orbit, they know that they are watching an event that took place seventeen minutes ago. All they need do is subtract this duration from the time read off a clock. If the clock reads 9:52, simple subtraction gives a result of 9:35.

Now suppose that at exactly the same time that this observation is made, a space vehicle is traveling past the earth, toward the comet, at a velocity equal to 70 percent of the speed of light. Suppose that some astronomers who are traveling on this vehicle make exactly the same calculation. What time will they obtain as a result?

Simple common sense tells us that they will also conclude that the comet passed the Martian orbit at 9:35. However, common sense can sometimes be misleading. In this case its application will give an incorrect result. The astronomers on the space vehicle will conclude that the comet passed the orbit of Mars not at 9:35, but at around 9:40. Even though they make the same kinds of measurements and perform the same calculation, their determination of the time of this event will not agree with that obtained by the earthbound astronomers.

Common sense tells us that if there is a discrepancy, then one set of observers must be right and the other set wrong. Again, common sense is misleading. There is no reason whatsoever for preferring one set of observations to the other; they are equally valid. According to Albert Einstein's special theory of relativity, time measurements made by observers in different states of motion will not agree with one another. In other words, time is relative; it is simply not possible to define the exact time at which a distant event took place. In fact, it is perfectly possible for a distant event to take place in the "past" of one observer and in the "future" of another.

Consider, for example, the following situation: Suppose that both sets of observers have made observations of the path of the comet long before it approached the orbit of Mars. Suppose they have concluded beforehand (while the comet is still, say, in the vicinity of the orbit of Jupiter) that its path will intersect the Martian orbit at 9:35 and at 9:40, respectively. Then, at 9:37, the observers on earth will conclude that this event has already happened,* while those in the spaceship will calculate that it will happen three minutes in the future.

Time in special relativity is not completely relativistic. Observers located near one another in space will always be able to agree on the meaning of the term "the present" when

* Of course, they will not actually see this happen until 9:52, since it takes seventeen minutes for light to travel from the orbit of Mars to earth.

that term is applied to nearby events. If a space vehicle is passing the earth just as a hydrogen bomb is detonated, the crew of the vehicle and earthbound observers will find it easy to agree that they have seen the blast at the same time. And if there were no event that both sets of observers could watch, they could still signal one another and use the signals to synchronize their clocks. In other words, they share the same "now." But this is not true of distant events. In relativity, the idea of "now" does not extend beyond "here."

These sound like bizarre and outrageous assertions. It might be best, therefore, if I put off further discussion of the implications of the special theory of relativity until later, and explain first what the theory is, and how it was discovered.

"Relativity" is not a concept that was invented by Einstein. In fact, Newtonian mechanics is also relativistic, although in a more restricted sense than the theories of Einstein. Newtonian mechanics satisfies a principle known as *Galilean relativity.* Newton's laws of motion imply that there should be no way to detect absolute motion. All that one can observe is the movement of two bodies relative to one another.

The principle is called Galilean relativity because it was Galileo who first pointed out that there are no experiments that can be performed on an oceangoing ship that will enable one to determine whether the ship is in motion or not. For example, an object that is dropped from a mast and allowed to fall to the deck will always seem to travel straight down, whether or not the ship is moving with respect to the earth, or to the water. If, for example, the ship is moving at a speed of ten knots, the object will be moving with the ship at a horizontal speed of ten knots as it falls. To those who watch it from the deck of the ship, only its vertical motion will be observable.

Early in the twentieth century, Einstein observed that while Newton's laws of motion satisfied a principle of relativity, Maxwell's theory of electrodynamics did not. Maxwell's theory implied, for example, that an electric field was always created when a magnet was set in motion. No one doubted that this was true. By Einstein's time, this phenomenon had been confirmed many times; it was the principle upon which electrical generators were based. However, there was something very odd, to Einstein's way of thinking, about this particular theoretical prediction. If moving magnets generated electric fields while stationary ones did not, then it should be possible to detect absolute motion.

In practice, no one knew how to do this. If one constructed a device that was designed to detect these fields, the device would give the same reading whether it or the magnet was moved. Einstein suspected that contrary to what Maxwell's theory implied, the detection of absolute motion was impossible. So he set out to see what would happen if Maxwell's theory was recast in relativistic form.

Einstein found that if the theory was to be modified in this manner, it was necessary to make the assumption that the speed of light, as measured by any observer, was constant. According to Maxwell, light was made up of electromagnetic vibrations; thus an assumption concerning the velocity of light (and of other electromagnetic radiations, which travel at the same velocity) had implications for the behavior of magnets. The mathematical equations that constituted Maxwell's theory linked the two phenomena together.

The assumption of the constancy of the velocity of light sounds like an outrageous idea. If it is correct, then it makes no difference whether a source of light is stationary or whether it is approaching me at a high velocity. In either case, if I measure the velocity of the light it emits, I will obtain the same result: 186,000 miles per second. Furthermore, I should still obtain this same result if I am traveling rapidly toward the source. After all, if only relative motion matters, it shouldn't make any difference whether it is approaching me or I am approaching it.

Einstein knew very well that in general, velocities did not exhibit this kind of constancy. For example, if a train is traveling at a speed of fifty miles per hour, and if a passenger throws an object forward through a passenger coach at a velocity of thirty miles per hour, the object will travel at eighty miles per hour with respect to the earth. If a jet plane is moving at a velocity of six hundred miles per hour, a rocket it fires at a speed of five hundred miles per hour will streak toward its target at eleven hundred miles per hour. Not even sound waves exhibit the constancy that Einstein postulated for light. If I move forward to meet a sound wave that is traveling toward me, its velocity will seem to be greater than it will be if I remain stationary.

Einstein must have realized that this assumption defied common sense. On the other hand, he saw that the postulate of the constancy of the speed of light not only allowed him to extend the principle of relativity to electrodynamics, it also cleared up certain long-standing difficulties in physics.

When Maxwell published his theory in 1873, it had been known for more than a century that light was a wave phenomenon. Now, wave phenomena are vibrations. Sound is produced when the molecules of the air are made to vibrate. Ocean waves can be thought of as slow vibrations of the surface of the sea. Not surprisingly, nineteenth-century physicists concluded that light must be caused by vibrations in some medium. Hence they postulated that space was filled with a substance called the *ether*. It was the ether that carried light vibrations from the sun to the earth, and that was responsible for the transmission of light on the earth's surface as well. Presumably, the ether not only filled "empty" space, it was also intermixed with the earth's atmosphere.

There were numerous problems with this theory, however. Calculations indicated that if the ether was to carry the rapid vibrations of light, it had to be quite rigid. On the other hand, it had to be rarefied enough to give no detectable resistance to the motion of the earth as the latter moved in its orbit around the sun. But it was not easy to see how these characteristics could exist in the same substance. Furthermore, no one was able to find a way to detect the existence of the ether experimentally. When Einstein published his paper on relativity in 1905, all such attempts had failed.

Einstein eliminated all of these problems with one bold

stroke. He suggested in his 1905 paper that the reason no one had been able to detect the ether was that it did not exist. It should be noted that experimental confirmation of this hypothetical substance would have been fatal to Einstein's ideas. If an ether existed, there would be such a thing as absolute motion. Either one would be moving with respect to the ether, or one would be stationary. Furthermore, in such a case, light would not travel at a velocity of 186,000 miles per second with respect to a moving observer; it would move with that velocity with respect to the ether.

Although it seems to have been theoretical considerations rather than experimental results that led Einstein to make his assumption concerning the speed of light, it is worth mentioning a famous experiment that was performed nearly twenty years before Einstein's theory was published, in which an attempt was made to measure the velocity of light with respect to the ether. In 1887, the American scientists Albert Michelson and Edward Morley had constructed an apparatus that could compare the velocities of rays of light that were traveling in different directions with respect to the motion of the earth. Reasoning that this motion should give rise to an ether "wind" that seemed to be blowing past the earth, they concluded that light would seem to travel more slowly when it went "upstream" than it did when it had the ether wind at its back. To their surprise, they discovered that no significant ether wind could be detected. Although they did obtain a small positive result, this result was much smaller than expected, and could be attributed to experimental inaccuracies.

Since 1887, experiments similar to that of Michelson and Morley have been performed on numerous occasions, and direct measurements of the velocity of light have also been made. These experiments confirm Einstein's assumption that the velocity of light, as measured by any observer, is always constant.

At the beginning of this chapter I made a number of assertions about the nature of time in special relativity. Perhaps it is now time to go back and to consider this matter in somewhat more detail. Perhaps the best way to begin would be to consider an old "thought experiment" that was invented nearly eighty years ago. Let us imagine that two bolts of lightning strike two ends of a train that is traveling rapidly toward its destination. We will imagine, also, that these lightning bolts are seen by two observers, one who is standing next to the tracks, and one who is riding in the center of the train.

Let us suppose that the observer on the ground is positioned midway between the two points at which the lightning strikes, and that he sees the two flashes at the same instant. He will quite naturally conclude that the two bolts of lightning were simultaneous. He realizes that he is not seeing the lightning at the instant that it strikes. However, since the light from the two flashes travels equal distances in equal periods of time, he readily concludes that the lightning struck in the two places at the same time.

Now let us consider the observer at the midpoint of the train. Since he is traveling away from one flash of light and toward the other, he will see the lightning bolt that struck the front of the train a small fraction of a second earlier. The light from the bolt that struck the rear will take longer to reach him, because it has to "catch up" with him as he moves away from it.

Now, suppose we bring in the assumption that motion is relative. This assumption doesn't alter the above description one bit, because it is perfectly legitimate to consider what is happening from the viewpoint of the observer on the ground. The man on the train will see one flash first, whether motion is relative or not.

Nevertheless, the assumption of the relativity of motion does have certain important consequences. In particular, it is possible for the observer on the train to assume that it is he, not the observer on the ground, who is motionless. In any event, the man on the ground wasn't really stationary anyway. He was standing on an earth that was rotating on its axis and revolving around the sun.

If the man on the train assumes that his reference system

is the motionless one, then he must conclude that the two bolts of lightning did not strike the two ends of the train simultaneously. From his point of view, the light from the two flashes travels toward him at equal velocities. Since the distances traveled are the same, he must assume that lightning hit the front end of the train a short time before the other bolt struck the rear.

Thus the two assumptions of the constancy of the velocity of light and the relativity of motion lead immediately to the conclusion that there is no such thing as absolute simultaneity of events that are separated in space. Spatially separated events that seem to be simultaneous in the reference system of one observer will, in general, not appear to be simultaneous in the reference system of another.

This explains the source of the disagreement between different observers in the example that was given at the beginning of the chapter. The difficulties associated with stating exactly when the comet intersected the orbit of Mars are a consequence of the fact that the concept of "now" cannot be applied to an event that is taking place 306 million kilometers away. Because the orbit of Mars is separated from us in space, we cannot say what "time" it is there now, or what time it will be there after another seventeen minutes have passed on our clock. To speak of "now" in some other place is to invoke the concept of simultaneity of spatially separated events. Special relativity says that this procedure is invalid.

Naturally one can compute a time for a distant event. However, this time only has a limited validity; it will not be the same as the time computed by an observer in a different state of motion. Hence we find ourselves in a position of not being able to say what time it "really is" at the orbit of Mars, or on the sun, or on a distant star. When phenomena appear different to observers in different states of motion, there is no way of deciding which is "right," because all are equally correct.

According to the special theory of relativity, observers in

different states of motion will not even agree upon the time order of events in certain cases. Suppose that we add a third observer to our thought experiment. Imagine that this observer is at the midpoint of a train that is passing the first train in the opposite direction. From the viewpoint of such an observer, lightning will strike the rear of the first train before it strikes the front. He will see the two events take place in an order that is the reverse of that perceived by the observer on the other train.

We can summarize this result as follows: If we have two spatially separated events that we will call A and B, and observers in different states of motion, some observers may conclude that A happened first, while others will insist that it was B that preceded A. In such a case, it is even possible for an event to take place in the "past" of one observer and in the "future" of another.

Special relativity does not make the concepts of "past" and "future" completely arbitrary, however. Nor does it contradict ordinary notions of causality. No observer, in any state of motion, will ever see a nail being driven into a piece of wood before it was struck by a hammer. Nor will any observer see a basketball fall through a hoop before it was thrown by a player. Finally, every observer will agree that the light from distant stars falls onto the surface of the earth long after it was emitted. The time order of events is observer-dependent only when the two events are distant enough from one another that no signal traveling at the speed of light could possibly get from one to the other before the latter happened. In the case of the lightning flashes, for example, the very fact that one observer saw them as simultaneous implies that they could not be linked in this manner by a signal.

What special relativity does imply is that whatever time is, it is not a substance that "flows" at an even rate throughout the universe. The "time" at which a distant event takes place is dependent upon the state of motion of the observer. Or, as physicists say, each observer has his own *proper time.* In general, this proper time does not coincide with the proper times of observers who are moving with respect to him.

Since a distinction was made in the previous chapter between the time of physics and subjective time, it might be well to point out that special relativity does not really depend upon the latter. When we speak of "observers" it is not necessary to think of them as conscious beings. An observer could just as easily be an electronic device that recorded events on videotape. Nor does special relativity depend upon the idea of "now." I have used such terms as "past," "present," and "future" for the sake of clarity; it would be possible to discuss special relativity without making any use of them. For example, one could speak of time t_1, time t_2, and so on. Such a discussion would, however, have seemed a bit too abstract.

When the mathematical equations associated with special relativity are worked out in detail, it becomes apparent that time can behave in strange ways indeed. The theory predicts the existence of a *time dilation* effect. The nature of this phenomenon can best be explained by means of an example. Suppose that a spaceship is traveling away from the earth at a velocity that is 87 percent of the speed of light. From the viewpoint of an observer on earth, time on the ship will slow down by a factor of two. Clocks on the ship will seem to be running only half as fast as those on earth. Biological processes will be similarly affected. The crew of the ship will seem to age only six months while one year passes on earth.

Since special relativity admits only of the existence of relative motion, it is perfectly legitimate for the people on the ship to take the point of view that they are motionless and that it is the earth that is traveling away from them. If they do, they will see the same effect in reverse. From their point of view, it will be time on earth that is passing more slowly. When their clocks register a passage of time of one hour, those on earth will advance only thirty minutes.

Now suppose that the ship travels to a distant star, and

then returns to earth at the same velocity. Suppose that the entire journey takes twenty years, according to the ship's chronometers. How many years will have passed on the earth, ten or forty?*

This question gives rise to what is called the *twin paradox*. The name is a reference to the fact that it is possible to rephrase the question by imagining that one member of a set of twins travels on the ship, while the other remains on earth, and then ask, "Which twin will be older?" In spite of the name, the twin "paradox" is not a paradox at all. In fact the question has a perfectly straightforward answer: The twin on the ship will age only half as fast as the one who remains at home. If the journey takes twenty years of ship's time, forty years will have passed on earth at its end.

The reason that only one interpretation is valid is that when the ship travels to another star and then turns around and comes back, we no longer have a symmetrical situation. At some point, the ship had to decelerate, come to a stop, and then accelerate in the other direction. It is not legitimate to imagine that the firing of the spaceship's rockets caused the entire universe to come to a stop, and then accelerate in the other direction. Although it is perfectly permissible for the observers on the ship to make the initial assumption that it is the earth that is traveling away from them, they cannot assume that the earth has turned around and is rushing to meet them during the second leg of the journey. Motion at a constant velocity is relative, but acceleration is not. When the time dilation calculations are worked out in detail, it is found that twice as much time passes on earth from either viewpoint. As one would expect, one twin does turn out to be older than the other; he does not find that he is older and younger at the same time.

The solution of the twin paradox demonstrates that

* It doesn't make any difference whether the ship is traveling toward or away from the earth. Time dilation depends only upon relative velocity, not its direction.

time dilation is not an illusion; it is quite real. Furthermore, the magnitude of the dilation increases with relative speed. If the ship were to travel at a velocity of 99 percent of the speed of light, time would slow down by a factor of seven; 140 years would go by on earth while 20 passed on the ship. And if the ship traveled at 99.9 percent of the velocity of light, the time that would elapse on earth would be 447 years. According to special relativity, there is a sense in which time travel into the future is possible.

But what about travel into the past? According to a limerick once published in the British magazine *Punch,*

> There was a young lady named Bright,
> Who traveled much faster than light.
> She started one day
> In the relative way,
> And returned on the previous night.

In a way, this description is perfectly accurate. According to the special theory of relativity, if anything could travel faster than light, it would go backward in time. However, it is extremely unlikely that Ms. Bright or anyone else is ever going to go off and return on the previous night. Special relativity also implies that the speed of light is a limiting velocity. It can be approached, but it can never be reached.

Special relativity also predicts that rapidly moving objects will experience a mass increase. If an object is traveling at 87 percent of the speed of light, there will not only be a slowing down of time by a factor of two, the object will also become twice as heavy. If its velocity is increased still further, its mass will continue to grow larger. At the velocity of light, its mass would become infinite. Since, as the mass grew, larger and larger quantities of energy would be required to accelerate the object, an infinite amount of energy would be required. Velocities equal to that of light are impossible.

Like the time dilation, the mass increase is dependent upon the state of motion of the observer. The crew of a spaceship traveling at a high velocity would not notice any increase in mass, even though observers on earth would. However, like the time dilation, the mass increase is quite real; it is not an illusion that is created by difference in velocity. If a grain of dust could be accelerated to a velocity close enough to that of light, it could acquire a mass equal to that of a planet. If such a dust grain struck the earth, the impact would be so great that our entire planet would burst into fragments.

I sometimes hear it said that there is much that is yet to be discovered and that, at some point in the future, we may gain knowledge that will show us how this "light barrier" can be broken. While it is true that numerous discoveries remain to be made, I think that it is unlikely that we will ever discover that faster-than-light speed is possible. New theories in physics do not often completely supplant old ones; more often, they simply improve upon them, or describe the behavior of matter under more extreme conditions. In other words, they typically do no more than provide us with better approximations.

When Einstein published his special theory of relativity, Newtonian physics was not discarded. In fact, one of the things that Einstein's theory confirmed was that Newton's laws of motion remained valid in cases where velocities were small compared with that of light. Newtonian mechanics is still used in cases where velocities are not large. The differences between its predictions and those of special relativity are so small that in most cases they cannot be measured.

One sometimes hears mention of hypothetical faster-than-light particles called *tachyons*. Special relativity does not deny that such particles could exist. In fact, it seems to predict their existence in a way that is similar to electrodynamics' prediction of time-reversed waves. The only restriction that special relativity seems to place on tachyons' behavior is that it implies that they can never pass the "light barrier" from the other side. If there are particles that travel faster than light, they can never decelerate to a velocity that is less than that of light.

At present, there is no evidence for the existence of tachyons. They are nothing more than an interesting theoretical possibility. It is possible that they are real, but that they do not interact with ordinary slower-than-light matter. But if this is the case, we can probably consider them to be nonexistent for all practical purposes. After all, something that cannot be seen can hardly be said to be "real" in the same sense as the objects we observe around us.

If tachyons did exist, and if they could somehow be used to transmit signals, then it would be possible to send messages into the past. As I noted previously, special relativity implies that anything that can travel faster than light could go backward in time. If such backward-in-time messages were possible, the world would be a paradoxical place indeed; it would be possible for a future event to "cause" a past one.

Tachyons, by the way, have nothing to do with the idea that a positron might be an electron that is moving backward in time. Like all other known forms of matter, positrons always travel at velocities less than that of light. And even though the backward-in-time theory is a possible interpretation of positrons' behavior, these particles cannot be used to send backward-in-time messages. If one were able, for example, to use positrons to transmit a message in Morse code, it would only be sent in the usual direction: toward the future.

Special relativity makes yet another important prediction concerning the behavior of objects that travel at high velocities. According to the theory, a rapidly moving object will exhibit a *length contraction*. It will become shorter along the direction of its motion. Thus, from the standpoint of an observer on earth, a spaceship that is traveling away from the earth will appear to shorten. From the viewpoint of an observer on the ship, the earth will flatten out.

It is the length contraction that causes disagreement between different observers concerning the times at which distant events take place. In order to see why this should be so, let us return once again to the example of the comet that passes the orbit of Mars. When astronomers on earth attempt to compute the "time" at which this happens, they do so by dividing the distance to the Martian orbit by the velocity of light. If they know the distance, and if they know how fast light travels, then they can easily find the time required for a ray of light to cross that distance.

But the length contraction ensures that astronomers on a spaceship that is traveling past the earth at a large velocity will obtain a different result. To these observers, the distance between the earth and the Martian orbit will be much less. The lentgh contraction not only affects material bodies, it also causes distances to be measured differently. Thus when these observers compute the time required for a ray of light to reach them, they will obtain a different result. Naturally, they could use a distance given them by earth astronomers if they so wished. But if they did, they would be computing a time in the earth's reference system, not a time in their own.

Both distances in space and intervals in time seem to be different to observers in different states of motion. And yet special relativity is based on the assumption that a quantity involving both space and time, the velocity of light, is always a constant. This seems to suggest that perhaps space and time are not entirely distinct entities, that there is a sense in which they are bound up with one another. In fact, physicists often find it convenient to replace the three dimensions of space and the single dimension of time with a four-dimensional geometry called *space-time*.

Space-time is not a concept that was invented by Einstein. It was developed by the Russian-German mathematician Herman Minkowski in 1907, two years after the special theory of relativity was published. Minkowski noted that neither time intervals nor spatial distances were invariant in relativity, but that it was possible to combine space and time in a mathematical way so that space-time intervals remained constant for any observer. Although observers in different states of motion will, in general, disagree about the times at which events take place, and about how widely separated they are in space, they will have no difficulty agreeing that the events have a given separation in space-time.

The use of the concept of space-time has created so many misconceptions among lay people that it might be best to begin by explaining not what it is but what it is not. First, the use of the space-time idea does not imply that time is a kind of space. In special relativity, time has basically the same character that it has in classical physics and in everyday life. It is still one-dimensional, and it still has arrows pointing in a certain direction.

The use of the space-time idea does not imply that there are four spatial dimensions. In fact, even though one can speak of a *four-dimensional space-time continuum,* neither Einstein nor Minkowski added a dimension to the usual three. For that matter, Newtonian mechanics is just as four-dimensional as relativity is. It, too, makes use of one dimension of time and three of space. The only difference is that in Newtonian physics, space and time do not interact with one another as they do in relativity.

The concept of space-time is used for only one reason: It makes the special theory of relativity mathematically simpler. It is perfectly possible to do without the concept, but it creates needless difficulties. Four-dimensional mathematics, after all, is not especially complicated; using it is simply a matter of introducing four coordinates in place of the usual three. Mathematically, a dimension is only a coordinate. Sometimes it is convenient to use three coordinates to describe a phenomenon, and sometimes it is easier to use four or more.

Mathematicians frequently deal with many-dimensional and even infinite-dimensional spaces without encountering insurmountable problems. The infinite-dimensional spaces are even used in physics sometimes, when physicists want to look at a problem in an abstract mathematical way. Naturally this does not imply that there are an infinite number of dimensions of space.

If there are misconceptions about space-time, there exist even more about the special theory of relativity itself. Special relativity is often viewed by lay people as an especially abstruse theory. In fact, not many decades have passed since the press would habitually print statements to the effect that there were only four (or five, or seven) men in the world who understood relativity (and apparently women were thought incapable of understanding it at all).

In fact, special relativity is one of the mathematically more straightforward theories in physics. It is simple enough that it can easily be taught to undergraduate students. Although its predictions sometimes seem bizarre, the basic concepts are really not very difficult to grasp. Einstein's general theory of relativity (to be discussed in the next chapter) is mathematically very complicated, but the special theory is not.

Nor does there seem to be much danger that the special theory will suddenly be overthrown. As a matter of fact, it is one of the best-confirmed theories in physics. Its predictions are tested every day, whenever physicists accelerate subatomic particles in modern particle accelerators. Whenever this is done, the particles exhibit mass increases and time dilations. Experiments have also been performed which demonstrate that macroscopic objects exhibit the predicted relativistic effects also. For example, the existence of time dilation was confirmed in 1976 when an accurate atomic clock was flown around a long racetrack in an airplane. Even though the plane flew at a relatively slow speed—compared to the velocity of light—the clock proved to be accurate enough to make the velocity-induced time dilation measurable. Another experiment, performed sixteen years earlier, had confirmed the existence of time dilation in a somewhat less direct manner when radiation emitted by a sample of iron-57 that had been placed in a centrifuge was compared to that emitted by a sample that was not spinning.

Since special relativity is so well established a theory, it appears that we must take its implications about the nature of time quite seriously, even though they seem to be rather bizarre at first. Since some of these ideas do seem strange to one who is used to thinking of time in commonsense, prerelativistic ways, it might not be a bad idea to summarize the contents of this chapter:

1. The "time" at which a distant event takes place cannot be defined in an unambiguous way. Since different times will be computed by different observers, the concept is really inapplicable. If we look at time in subjective terms, we can say that "now" does not extend beyond "here."

2. Rapidly moving objects exhibit a time dilation effect. This effect is real, not illusionary, as the twin "paradox" demonstrates.

3. If two events are so close in time or so widely separated in space that no signal traveling at the speed of light can possibly get from one to the other before the latter takes place, then their time order is ambiguous. Some observers will conclude that event A happens first, others that event B took place earlier in time.

4. Although we can loosely say that special relativity implies that "time is relative," the theory does not contradict ordinary ideas of causality. No observer will ever see a basketball fall through a hoop before the player who has thrown it makes his shot.

CHAPTER 10

Cosmic Time

DURING THE EARLY YEARS OF the nineteenth century, the German mathematician Karl Friedrich Gauss spent some time making a survey of the kingdom of Hanover. In 1827 he wrote a paper in which he recorded measurements made of a triangle formed by three mountain peaks, Brocken, Hohehagen, and Inselberg. Gauss had been trying to determine whether the sum of the three angles of the triangle was 180°, or somewhat less.

Now, every high school student who studies geometry learns that the sum of the angles of *every* triangle is 180°. There is a theorem that proves that this is always true; there can be no exceptions. So what, exactly, was Gauss doing? Was he questioning the validity of geometrical theorems that had been accepted as conclusive since the time of the ancient Greeks?

Not exactly. Gauss was trying to determine whether or not space was curved. He was asking whether the geometry of space was really the traditional *Euclidian geometry,* or whether a *non-Euclidian* geometry was applicable. Only in Euclidian geometry is the sum of the angles of a triangle equal to 180°. In a non-Euclidian geometry, it can be more or less.

Gauss' experiment was not very conclusive. His measurements gave a sum that exceeded 180° by about 15 minutes of arc, or approximately four one-thousandths of a degree. But since the experimental inaccuracies were much larger than this, nothing was demonstrated. Gauss realized that the true sum of the angles could be 180°, or a little more, or a little less. All three conclusions were consistent with his data. He was unable to determine whether the geometry of space near the surface of the earth was Euclidian or not.

Euclidian geometry is the geometry that was systematized by the Greek mathematician Euclid around 300 B.C. Euclid based his geometry on a number of axioms and postulates, all of which he regarded as self-evident. For example, one of the axioms says that equals added to equals give equals; one of the postulates states that a straight line is the shortest path between two points. The axioms and postulates are the principles from which the theorems of Euclidian geometry are derived.

One of the postulates must have seemed to Euclid to be somewhat less self-evident than the others, for he proved as many theorems as he could before he made any use of it. This is the so-called *parallel postulate,* which states that one and only one parallel can be drawn to a line through any given point outside that line.*

After the death of Euclid, Greek mathematicians, who also found the parallel postulate a little suspect, attempted to replace it with another postulate that seemed more intuitively obvious, or to find a way to derive it from the other postulates and axioms. But all these attempts failed. Although the parallel postulate seemed somewhat questionable, no one could find a way to do without it.

At the beginning of the nineteenth century, such attempts were still being made. However, they were no more successful than those of the Greeks. If there was a way to re-

* This is actually a modern version of the postulate. Euclid himself stated it in a somewhat more complicated way.

state the parallel postulate in a really self-evident form, or to derive it, no one could find it. Finally, a few mathematicians began to wonder what would happen if the parallel postulate was replaced by something else. Suppose one assumed that *no* parallels could be drawn through a point exterior to a line? Suppose a line had many parallels? Was it possible to base a geometry on such strange assumptions? Would such a geometry be consistent (i.e., contain no contradictions)?

These questions were pondered by Gauss, by the Russian mathematician Nikolai Ivanovich Lobatchevsky, by the Hungarian mathematician Johann Bolyai, and by the German mathematician Georg Friedrich Riemann. They found that the answer to these questions was yes, that other kinds of geometry were possible.*

Non-Euclidian geometrics differ from the Euclidian variety in a number of different respects. In the geometry of Gauss, Bolyai, and Lobatchevsky, which is based on the assumption that there is more than one parallel through any given point, the angles of a triangle are less than 180°, and the areas of geometrical figures such as triangles and circles are given by different formulas from those obtained in Euclidian geometry (for example, the area of a circle is not πr^2). Riemann's geometry, which is based on the idea that there are no parallel lines, seems even stranger. In it, any number of distinct straight lines can be drawn through two points, and there is no such thing as a line of infinite length.

A physical space that is described by a non-Euclidian geometry is called a *curved space*. Although this term is commonly used by scientists, who give it a precise mathematical meaning, it can be somewhat misleading to lay people. It should be understood that the fact that a three-dimensional space is "curved" does not imply that there is a fourth dimension that the space is "curved in." Nor is it necessary to think of space as an object that is deformed. In fact, it is simpler to

* Since Gauss failed to publish his results, it is Lobatchevsky, Bolyai and Riemann who are generally credited with the discovery of non-Euclidian geometry.

think of curved space as a space that has a particular kind of geometry than as something that is physically warped.

If a three-dimensional space has a geometry like the one investigated by Gauss, Lobatchevsky, and Bolyai, it is said to have *negative curvature*. A negatively curved space that has no boundaries is infinite in extent, just as a *flat* (Euclidian) space is. If the space of our universe is negatively curved or flat, a ray of light emitted in any direction will continue on forever.

On the other hand, a positively curved space is finite. This follows from the fact that in this kind of geometry there are no lines of infinite length. If the space of our universe is positively curved, then a ray of light emitted in any direction will eventually reach its starting point from the other direction (provided, of course, that the universe lasts long enough for it to make a complete circuit). The path of such a light ray would be roughly analogous to that of an airplane that flew a complete circuit around the circumference of the earth.

If our universe has a positive spatial curvature, then it must be finite. However, this does not imply that it has any boundary. One could never reach the "edge" of such a universe, for it has no edges. Nor would it be possible to travel "outside" of it; the term "outside" has no meaning.

Again, the surface of the earth provides an analogy. But there is a point at which the analogy breaks down. The earth has a finite two-dimensional surface that is curved in a third dimension. As a result, it is possible to get "outside" of the two-dimensional surface by flying upward in a rocket or an airplane. Such an option would not exist in a positively curved three-dimensional space.

Although Gauss seems to have pondered the question of whether space was curved or flat early in the nineteenth century, nearly a century was to pass before anyone made a serious attempt to follow up his speculations. Physicists and mathematicians alike assumed that non-Euclidian geometries were only mathematical curiosities. Few doubted that it was the flat, Euclidian geometry that described the space of the physical universe.

This situation was suddenly changed in 1915 when Einstein propounded his general theory of relativity. According to Einstein's theory, space was curved by the presence of gravitational masses. Gravity, Einstein said, existed because the presence of mass gave space a non-Euclidian geometry.

Strictly speaking, general relativity deals not with curved space but with a curved, four-dimensional space-time. Although the concept of space-time can be dispensed with in special relativity if one so desires, the equations of the general theory indicate that the space and time coordinates will interact with one another in a more complicated way, and a four-dimensional mathematical formulation is essential.

It is easy to see why this should be the case. We need only recall that in special relativity, neither distances nor times remain constant when viewed by different observers traveling at different velocities. On the other hand, the velocity of light, which involves both distance and time (light travels at 186,000 *miles* per *second*) is always the same. General relativity, on the other hand, concerns itself not with relative velocities but with observers who are accelerated with respect to one another. Since time enters twice into acceleration (for example, all objects near the surface of the earth accelerate at the rate of thirty-two feet per *second* per *second*; every second, they move thirty-two feet per second faster), the general theory of relativity relates time to space in a rather complex way.

One of the postulates upon which general relativity is based is the *principle of equivalence.* This is really quite a simple idea. It had been known since the time of Galileo that an obvious characteristic of gravity was that it caused things to accelerate. Falling objects did not move toward the ground at a constant velocity; they went faster and faster. Einstein realized that this implied that gravity and acceleration were somehow equivalent. He saw that if he could express this equivalence in a mathematical form, he could relate both gravity and acceleration to the curvature of space-time, and obtain a theory of gravity.

There is a simple example that can be used to illustrate the principle of equivalence. Suppose that I am given an anesthetic and that when I wake up I am in a completely enclosed room. Suppose, further, that I feel a normal pull of gravity. Can I conclude that the room is part of a building on earth? Or is it also possible that the room is a cabin in a spaceship that is far from earth, and which is traveling at an acceleration of one g?

The answer is that I can't tell the difference. The earth's gravitational pull and a one-g acceleration would feel exactly the same to me. Nor is there any experiment that I could perform to distinguish between the two. The very use of the term "g" (for "gravities") to measure acceleration is a reflection of this.

General relativity is mathematically very complicated. However, the ideas upon which it is based are not. Perhaps the best way to illustrate this would be to give the basic tenets of general relativity in outline form, while restating the main points of the above discussion:

1. Special relativity deals with cases where observers move, relative to one another, at constant velocities.

2. General relativity describes what happens when observers accelerate with respect to one another.

3. But it is impossible for an observer to tell the difference between the effects of gravity and those of acceleration in space; the outstanding characteristic of gravity is that gravitating bodies cause objects to accelerate.

4. Therefore a theory which describes accelerated reference systems will also be a theory of gravity.

5. Special relativity predicts that there should be changes in times and lengths. The changes predicted by general relativity are more far-reaching. To an accelerated observer, the very geometry of space-time will seem to be different. The geometry of space-time is changed in a similar way by gravity.

In fact, according to the general theory of relativity,

gravity *is* the curvature of space-time produced by the presence of massive objects.

If curved space is hard to visualize, then it would appear that it would be practically impossible to form an intuitive idea of space-time. Fortunately, matters are not as difficult as they appear. The curved space-time of general relativity is really not so different from the curved space of three-dimensional non-Euclidian geometry. In fact, when the general theory of relativity is applied to the structure of our universe, one finds either a *negative* (the geometry of Gauss, Lobatchevsky, and Bolyai) or a *positive* (geometry of Riemann) curvature. It is possible to discuss the curvature of space and the "curvature" of time separately.

If the average curvature of space in the universe is negative, then the universe extends infinitely in every direction, and time is also infinite. In such a case, the universe would be said to be *open*. A negatively curved universe would go on expanding forever. If space is positively curved, then the universe is finite, and it also has a finite existence in time. Such a universe is said to be *closed*. In the case of a closed universe, general relativity implies that gravitational retardation will cause the expansion to stop eventually. The universe will contract into a smaller and smaller volume of space until it extinguishes itself in a *big crunch* that is analogous to the big bang.

General relativity does not tell us whether the universe is positively or negatively curved, or whether it has a finite or an infinite future. Einstein's theory tells us that either kind of universe is possible. This is a matter that must be settled by observation.

The average curvature of space depends upon the amount of matter that the universe contains. If the average mass density is more than a certain critical figure, called, naturally, the *critical density*, then the universe is closed. The critical density can be easily calculated. It is 5×10^{-27}*

* 10^{-27} is 1 divided by 10^{27}.

kilograms per cubic meter, or approximately three hydrogen atoms per cubic yard.

As we saw in Chapter 6, astronomers have not been able to determine how much matter the universe contains. There are certain theoretical arguments that seem to imply that the amount of matter in the universe is no more than 10 percent of the critical density. But this evidence is somewhat less than conclusive; astronomers are aware that these arguments may contain enormous holes.

The problem of determining the amount of time that has passed since the big bang and the question of whether the universe is open or closed are related. If astronomers could determine the mass density of the universe, both would be solved. Similarly, a determination of the rate at which the expansion is slowing down would provide answers to both questions. If one knew the deceleration, both the mass density and the age of the universe could be calculated.

Incidentally, the concept of the age of the universe does have a precise meaning, even though the special theory of relativity implies that time measurements are relative. General relativity is, paradoxically, less relativistic than the special theory. Although it doesn't contradict what special relativity says about time measurements made by different observers, it does allow one to define a *cosmic time* that can be applied to the universe as a whole.

Cosmic time is the time that would be measured by an observer who was moving along with the average expansion of the universe. Like the time of thermodynamics, cosmic time is a statistical concept. Individual stars and galaxies may have random velocities in this or that direction. But on the average, their motion will coincide with that of the universe. Things could hardly be otherwise, for the stars and galaxies, and other matter, are the universe.

The expansion of the universe causes galaxies to recede from one another. If they are very far apart, this velocity of recession can be a significant fraction of the speed of light. Nevertheless, observers on the various different galaxies will measure the same cosmic time.

For all practical purposes, cosmic time is identical with earth time. It is true that the earth is rotating on its axis and revolving around the sun, and that the sun is revolving in an orbit around the center of our galaxy. However, all of these velocities are small compared to that of light. Therefore relativistic time dilation effects will be small, and earth time will coincide with cosmic time to a very good approximation.

Consequently, when we say that the universe is 10 or 15 or 18 billion years old, it is possible to take these years to be earth years. Of course, the earth itself is less than 5 billion years old. But this really isn't very relevant, since we're only taking its year as a standard of measurement, not implying that it has existed as long as the universe itself.

However, it would be a mistake to assume that all objects in the universe share a universal time that is roughly equivalent to earth time. The effects that gravitational masses have on the geometry of space cannot be separated from the effects that they have on the geometry of time. In particular, gravitational fields produce a time dilation effect that is quite similar to that encountered in special relativity. All gravitational fields slow down the passage of time, and very intense gravity causes a dilation that is quite dramatic. In some cases, gravity causes time to come to a complete stop, at least from the viewpoint of an observer who is located at a distance from the fields that cause the time dilation.

The objects that cause time to come to a halt are black holes, collapsed remnants of massive stars that have gravitational fields that are so intense that nothing, not even light, can escape from them. At present, astronomers do not have any direct observational evidence that black holes really exist. However, there is circumstantial evidence for their existence that is reasonably convincing. Furthermore, the general theory of relativity seems to predict that black holes must form whenever a star that has more than a certain critical mass collapses.

The outer layers of a star are held up by the pressure created by the nuclear reactions that take place within the star's interior. But the nuclear fuel that feeds these reactions can last only for a certain period of time. Furthermore, the larger a star, the shorter the time that its fuel will last. The greater quantity of matter within a large star is more than compensated for by the fact that the nuclear fuel will be consumed at a much faster rate.

Stars the size of our sun have lifetimes of about 10 billion years. Stars that are less massive can live many times longer than that, while very large stars may have lifetimes of a few million years or less. Since the universe is much more than a few million years old, there has obviously been ample time for such stars to be created, use up their fuel, and die.

Not all dying stars collapse into black holes. Our sun, for example, will never suffer this fate. It will evolve into a *red giant* as it nears the end of its life about 5 billion years from now, and then it will gradually contract into a cool (cool by stellar standards, that is), dim *white dwarf*. Stars that possess more than about 1.4 solar masses at the ends of their lives experience a different fate. These stars contract into *neutron stars*. Neutron stars derive their name from the fact that when gravitational forces exceed a certain limit, the remnants of the star will become so compressed that protons and electrons are squeezed together and neutrons are formed in their place. A neutron star can be thought of as an object that is a massive neutron "sea."

White dwarfs are very dense objects. It has been estimated that a matchbox full of white dwarf matter would weigh about ten tons in the earth's gravitational field, and that a measuring cup full would weigh more than two dozen elephants. Neutron stars are even denser. The matter at their centers is about 10^{15} times as dense as water. A cubic centimeter of such matter would have a mass of about a billion tons.

The evidence for the existence of white dwarfs and neutron stars is quite convincing. White dwarfs can be observed through telescopes, and measurements of their brightness make it possible to calculate their mass and density. Neutron stars can be observed also, because they emit pulses of radiation in the form of radio waves, light, or X rays. When they are components of double star systems, estimates of their masses can be obtained from studies of their orbital motion. Calculations of neutron stars' density are based on theoretical ideas about their structure. However, there is little reason to doubt these theoretical results.

Currently accepted theory indicates that there is a limit to the amount of mass that a neutron star can have. It is estimated that this limit is somewhere between 1.6 and 3.0 solar masses. If a neutron star is heavier than this, then gravitational forces should become so strong that the matter of which the star is composed will not be able to resist further compression.

Once this process of compression begins, there is nothing that can stop it. The matter in the star may resist further collapse, but the pressure created by this resistance will only make the collapse proceed faster. In general relativity, the existence of pressure only makes gravitational forces stronger. The matter in the star will thus be squeezed together until it occupies zero volume and the density of matter becomes infinite. Such a point of infinite mass density is called a *singularity*.

Although physicists and astronomers have a great deal of confidence in the predictions of general relativity, many of them are somewhat skeptical of this one. They doubt that there really are such things as singularities. They point out that the existence of infinities in theoretical calculations is generally an indication that the theory has broken down. In their view, general relativity's prediction of infinite densities is a sign that the theory has been pushed beyond the limits within which it is valid.

This argument sounds quite reasonable. One would not expect that general relativity should continue to be applicable when all the matter in a massive star is compressed into a region much smaller than an atomic nucleus. To describe gravitational events that take place within such a small volume, general relativity would have to be combined with quantum mechanics, the theory that is used to describe events that take place on the subatomic level.

Physicists have a name for a theory that would combine general relativity with quantum mechanics. Such a theory would be called a theory of *quantum gravity*. Unfortunately, no one knows how to go about constructing such a theory; this is one of the unsolved problems of theoretical physics. Consequently, no one really has any idea whether singularities exist or not.

But even if they do not, it is likely that the matter of the star is compressed into an incredibly small volume. It has been estimated that gravity should be modified by quantum effects only at distances of 10^{-33} centimeters or less. This is many orders of magnitude smaller than an atomic nucleus, which has dimensions of the order of 10^{-13} centimeters. Even if such a compressed remnant of a star could be observed, scientists might not be able to tell whether it was a point, or a region of small but finite size.

And, of course, a singularity (from this point on, I will speak of singularities as though they exist, with the understanding that the above reservations are being kept in mind) could not be observed, even if it were possible to travel to a black hole. It is not possible to see into a black hole to view the singularity at its center, or anything else that is inside a surface called the *event horizon*. Anything that crosses the event horizon and enters the black hole is trapped within the horizon forever. Not even light can escape.

The event horizon should not be thought of as a physical surface. It is simply the place at which the gravity created by the matter in the singularity becomes so strong that anything that crosses the horizon is trapped. The event horizon is a sphere that has a diameter approximately equal to six kilometers multiplied by the mass of the black hole in solar masses. Thus a black hole three times as heavy as the sun would be about eighteen kilometers (eleven miles) across. A black hole twenty times as heavy as the sun would have a diameter of about 120 kilometers (or a little more than seventy miles).

The event horizon is also the surface at which time comes to a stop, at least from the standpoint of an outside observer. In order to see precisely what this means, let us imagine that a space vehicle is approaching a black hole. From the viewpoint of an observer who is situated some distance away, time on the ship will appear to pass more and more slowly as it approaches the event horizon. When the ship reaches the horizon, time will stop, and the ship will remain suspended there for all eternity. In other words, to observers in the outside universe, the ship will never enter the black hole.

But things will have an entirely different appearance to observers on the ship. Time will continue to pass in a normal manner, and there will be nothing to prevent them from traveling across the event horizon, into the black hole's interior. If the black hole is one that has a mass that is several times greater than that of the sun, they will never have a chance to observe anything that is going on inside, for intense gravitational forces will rip the ship apart shortly after the event horizon is passed, if indeed this has not already happened before the black hole is entered.

But there is nothing to prevent us from imagining that the ship has entered the event horizon of a black hole of, say, a million solar masses. Many astronomers believe that such supermassive black holes may exist at the centers of galaxies. There is so much matter in galactic cores that black holes of such a size could easily be created. Once a black hole did form in such a region, it would grow in size quite rapidly, since its gravitational fields would capture enormous quantities of interstellar gas and dust, and gobble up entire stars as well.

A black hole of a million solar masses would have a diameter of about 6 million kilometers. A ship that passed through the event horizon would be far enough from the singularity that its occupants could manage to explore the region inside, at least for a time. Of course, they would still be committing suicide, for once inside the black hole, they would be drawn toward the singularity, and what remained of their ship would be crushed out of existence when the singularity was reached.

Not only would observers on the ship observe events

that were taking place within the black hole, they would also be able to see events that were taking place in the outside universe. Light cannot escape from a black hole, but of course it can cross the event horizon from the outside. The occupants of the ship would even be able to receive messages that had been sent to them from the oustide universe. However, one thing that they could not do would be to observe the singularity at the center. Light and other forms of electromagnetic radiation can travel toward a singularity, but not away from it. The singularity would be as invisible to observers on the ship as events within the black hole would be to observers outside.

It seems incredible that from the viewpoint of one observer, time should grind to a halt, leaving the ship frozen on the event horizon, while time should continue to pass in a perfectly normal manner from the viewpoint of other observers. But there is really nothing contradictory about this conclusion. The ship would not be in two places at the same time. In fact, the term "same time" doesn't even have any meaning in this context. In such a situation, there would be two different times: time on the ship and time in the outside universe. When the ship reached the event horizon, the discrepancy between the two would become so great that they could not longer be matched up. On the other hand, the behavior of the ship in either reference frame would be perfectly unambiguous. At any given time, it would be in a specific place.

The gravitational time dilation at the event horizon of a black hole could have amusing effects. Let us suppose, for example, that one of the astronauts had previously been an operatic tenor, and that he was singing an aria (perhaps from Wagner's *Götterdämmerung*) as the event horizon was approached. Then, to observers in the outside universe, it would seem that he was holding the same note for all eternity. Suppose, for example, that it was an E flat. In that case, the tenor would be singing that E flat when the event horizon was reached, and he would still be singing the E flat when the sun became a red giant 5 billion years from now.*

Or at least this is what the general theory of relativity predicts would happen. But how much confidence can we have in the theory? Can we really be sure that black holes exist? If they do, can we be sure that general relativity correctly describes the behavior of objects that approach or enter them?

In order to answer these questions, it is necessary to make observations and to perform experiments. Theories, after all, must be tested. There have been numerous plausible theories that have turned out to be incorrect, and even Einstein sometimes made mistakes.

Perhaps it would be best if we were to take the questions one at a time. Are there really black holes in the universe? Probably. Although black holes cannot be seen, it is possible to obtain evidence by which their existence may be inferred indirectly. If a black hole is part of a double star system, and if the black hole and its companion orbit one another at reasonably close distances, the gravitational fields of the black hole will draw matter off the surface of its companion. As this matter—mostly hydrogen, since hydrogen is the primary constituent of stars—is drawn toward the black hole, it will be accerelated. The acceleration will cause the matter to emit radiation in the form of X rays. Naturally, any X rays that are given off after the material reaches the event horizon will not be seen. However, copious amounts will be emitted before this happens.

Astronomers have observed numerous X-ray sources in the universe. Naturally, observations of X rays do not automatically indicate that a black hole has been discovered. There are a number of other ways in which this type of radiation can be produced. However, some X-ray sources flicker very rapidly. The X rays vary in intensity over periods of a

* Admittedly, there is one little flaw in all this. The time dilation would slow down the sound vibrations so much that the note could no longer be heard. But perhaps we can disregard this so as not to spoil the joke.

few thousandths of a second. This flickering is evidence that the X-ray-emitting region is very small; a large object would emit radiation at a fairly steady rate, because variations in intensity would average out.

But the observation of flickering does not provide conclusive evidence that a black hole has been observed. Other kinds of confirmation are necessary before any conclusions can be reached. In particular, it must be established that the flickering X rays are coming from a double star system, and that one of the objects in the system is too dark to be seen. But even after this is done, one can still not be certain that a black hole has been found. After all, the invisible object may not be a black hole at all; it might simply be an ordinary star that is too dim to be observable. It is necessary to find a way to calculate the mass of the invisible object in the X-ray-emitting region. Only if this object turns out to be very massive is there any reason to suspect that it may be a black hole.

If astronomers could make direct observations of the orbital motion of the visible component of such a double star system, it would be easy to calculate the mass of the star's companion. Unfortunately, most stars are so distant from the earth that it is not possible to do this. However, the motion of a star can be inferred from a study of the light that it emits. When its orbital motion causes it to move away from the earth, its light will be red-shifted. When it swings around and begins to move in the other direction, the light will be blue-shifted instead. Studies of such red shifts and blue shifts indicate that some double star systems contain dark components that are massive indeed.

At present, the best black hole candidate is Cygnus X-1, an X-ray source in the constellation Cygnus that was discovered in 1965. It has been established that the X rays come from a binary system that has two components: a bright blue giant star that is about twenty-five times as massive as the sun, and a dark object of ten to fifteen solar masses. The dark object is believed to be a black hole because, according to general relativity, any dark, compact object of this mass *must* be a black hole; there are no other possibilities.

Are there really black holes in the universe? If it were possible to actually observe a black hole collapse, or to send a spaceship out to Cygnus X-1 to observe it more closely, it might be easy to answer this question. But since it is not possible to do this, it seems that what it all boils down to is the question "How much confidence can we have in the predictions of general relativity?"

It is not so easy to answer this question either, for competing theories of gravity do exist, and the experimental evidence for the validity of general relativity is much less convincing than the evidence available to confirm the special theory. It is difficult to test general relativity because very intense gravitational fields are not available to us; experiments must be performed in the relatively weak gravity that exists in the vicinities of the sun and of the earth.

However, various kinds of experimental confirmation have been obtained. The development of accurate atomic clocks has made it possible to measure the time dilation caused by the earth's gravity. In order to accomplish this, it has been necessary to perform experimental *tours de force.* For example, one experimental confirmation of general relativity depended upon the measurement of a frequency change in some observed radiation that was only present to the extent of one part in 10^{15}. Nevertheless, numerous ingenious experiments have been devised, and within the limits of accuracy that are possible, they confirm general relativity's predictions.

It is also possible to test the predictions that general relativity makes concerning the gravitational effects of the mass contained in the sun. For example, the orbit of the planet Mercury exhibits a kind of perturbation called *precession of the perihelion* (Mercury's rather elongated orbit does not remain fixed; it "swivels" around the sun over a period of centuries) that seems perfectly in agreement with Einstein's predictions. It is also possible to measure the effects that the curvature of space near the sun will have on light or on other radiation that grazes the sun's surface. It is found that light rays that just graze the sun are deflected by the amount that general relativity says they should be. Radar beams have been bounced off Venus, Mercury, and Mars when those planets have been positioned on the opposite side of the sun from the earth. It has been found that when a radar beam grazes the sun, the sun's gravity will cause a slight time delay as the beams make their way back to earth. Again, the results are in accord with theoretical predictions.

In every case, the effects that have been measured have been minute. Thus it has not been conclusively demonstrated that general relativity is really better than certain of its competitors. However, these competing theories are generally nothing more than modifications of Einstein's theory, and have the disadvantage that they are somewhat more complicated. Since physicists always choose the simplest theory that will explain a given set of phenomena, and turn to more complicated ones only when the simple theory no longer seems to work, general relativity is still universally accepted.

Obviously it is one thing to make measurements in the earth's weak gravitational field and quite another to extrapolate to the intense fields that would be present around a black hole. However, there seems to be no good reason why this cannot be done. Physicists have discovered nothing that would lead them to suspect that any modifications of general relativity might be necessary before the region of quantum gravity is reached. Although general relativity is not an easy theory to test, scientists have discovered nothing that would seem to contradict it in the seventy years that have passed since it was originally proposed.

It appears that even though the existence of black holes has not been conclusively demonstrated, we must conclude that they very probably do exist. Furthermore, there is every reason to believe that the predictions of general relativity concerning phenomena that would be observed near a black hole's event horizon are accurate. Unless general relativity is suddenly overthrown, we must continue to consider it very likely that there really are situations in which time can come to a stop.

CHAPTER 11

The Beginning and End of Time

As we have seen, astronomers are not sure exactly how old the universe is. Although an age of the order of 15 billion years is accepted by many of them, some maintain that it is younger than that. There is even less agreement on the question of whether the universe is open or closed. Until a few years ago, the preponderance of evidence seemed to point toward an open universe. But all that was changed by the discovery that galaxies were surrounded by dark haloes of unknown composition. As long as no one knows what these haloes are, it is impossible to estimate how much matter the universe contains.

And yet there is virtually universal agreement that the universe began in a big bang. At first glance, this seems somewhat paradoxical. If there is so much about the present state of the universe that is uncertain, how is it possible to understand the nature of an event that took place billions of years ago? For that matter, how can one even be sure that there was a big bang?

Most astronomers would agree that it is possible to be reasonably certain that the universe did begin in this manner, for there is a sense in which the big bang can still be seen. Even though this explosive event took place something like 15 billion years ago, its traces are still visible today.

There are three important kinds of evidence for the big bang. First, the universe is expanding. Observations of distant galaxies indicate that this expansion has been going on for billions of years. Thus the universe must have been in a relatively compressed state at some time in the past. In fact, if one looks at the present expansion and works backward, it is hard to avoid the conclusion that the universe must have originally exploded out of a hot, dense fireball.

If this fireball existed, two important consequences follow. First, the light from the fireball should still be observable. Astronomers know of nothing that would keep it from continuing to travel through space during the 15 billion or so years that have passed since it was emitted. Second, the existence of a fireball at one time implies that the universe should have a certain kind of chemical composition. It can be deduced that certain kinds of nuclear reactions would have taken place within the fireball. The products of these reactions should still exist in the universe today.

Both of these predictions of the big bang theory have been confirmed. It is possible to see the light from the creation event, and the universe does have the kind of composition that it should have. Since there are no other theories that are capable of explaining these observations in any reasonable kind of way, scientists have concluded that the universe must have originally been in a hot, highly compressed state, and that it began in a state of rapid expansion.

There is a sense in which astronomy differs from other physical sciences. When scientists perform experiments in laboratories, they examine phenomena that take place in the present. Astronomers, on the other hand, examine events that took place in the past. Even when they make observations of a nearby object like the sun, they never see the sun as it is at the moment of observation. Since it takes a little more than eight minutes for light to travel from the sun to the earth, one can observe the sun only as it was eight minutes previously.

When astronomers look at distant galaxies, they look

ever farther into past time. If the currently accepted distance scales are correct, the most distant galaxies that can be observed with modern telescopes are more than 10 billion light-years away. Thus, if the universe really is about 15 billion years old, then observing these galaxies allows one to see events that took place about 5 billion years after the beginning.

Observations of the radio waves that fall on the earth from all directions of space allow one to see an even earlier era, for these radio waves are the remnants of the light that was emitted in the primeval fireball. Discovered in 1964 by two Bell Telephone Laboratory scientists, Arno Penzias and Robert Wilson, these radio waves are called the *cosmic microwave background radiation.* The name is a long one, but its meaning is relatively straightforward. "Microwave radiation" is a name for radio waves with wavelengths of less than one meter. "Cosmic" is a reference to the origin of these waves, and "background" means that they seem to be everywhere. The radiation that Penzias and Wilson discovered does not seem to come from stars, or from galaxies, or from any other discrete sources. It is a "background" that comes from every direction of the sky. Furthermore, it is always the same; if one points a radio telescope at different regions of the sky, the intensity of the microwave background varies by less than three parts in ten thousand.

If there was a primeval fireball that emitted copious quantities of light, one would not expect to see visible light in the sky today. Light that has been traveling through space for 15 billion years should undergo an enormous red shift. The principle that governs this red shift is exactly the same as the principle that applies to the light emitted by distant galaxies. It doesn't make any difference whether light is emitted by a star or by events that took place in the fireball itself. As time passes, and as the light travels through space, wavelengths become longer in either case. Since the big bang is farther away in time than any visible star or galaxy, the light that it emitted should show red shifts that are quite dramatic.

The light should be transformed into radio waves, the form of electromagnetic radiation with the longest wavelengths. And, of course, this radiation should come from all directions, because all of the observable universe, including the earth, is made of material that was once inside the fireball.

In order to understand what astronomers are seeing when they measure the microwave background, it is necessary to look at the big bang theory in a little more detail, and to see just what should have been happening during the early stages of the expansion of the universe. In particular, we need to know whether the microwave background had its origin in processes that were taking place at the moment of creation, or whether it is a product of events that took place some time afterward.

The big bang theory is based on the assumption that the universe was originally very compressed and very hot. It must have been hot, because it would have cooled as it expanded. This is analogous to the cooling of gases as they undergo expansion. For example, gas that is released from an aerosol can feels cold because it expands into a larger volume as it passes through the nozzle.

It is possible to perform calculations that indicate how hot the universe must have been at any given time in its past. In particular, it should have had a temperature of about 3000 K a million years after the beginning. Here the letter K stands for Kelvins, or degrees above absolute zero. Since absolute zero is −273° C, Kelvins can be converted to Celsius degrees by subtracting 273. 3000 K would therefore be 2737° C. In practice 2700° C would be close enough, since we are dealing with approximate figures.

The cooling of the universe to 3000 K represented an important transition point in its evolution. While the temperature was still greater than this, matter could not have existed in the form of ordinary atoms. The heat was so intense that the atoms must have been *ionized;* the energy that was flowing through the universe at this time was so great that electrons were kicked out of atoms as soon as they were captured.

Now it so happens that light interacts quite strongly with free electrons. Sometimes the electrons absorb the light, and reemit it a small fraction of a second later. More often they scatter it, and cause the light to gain or lose energy in the process. If there had been any conscious observers present in the universe at this time, they would have been able to see nothing but a dense fog that was glowing with yellow light.

But then, when the universe was somewhat less than a million years old, the ongoing expansion caused the temperature to drop below 3000 K. Suddenly the free electrons were able to attach themselves to the atomic nuclei that were everywhere present, and atoms were formed. Since atoms scatter light to a much lesser extent than free electrons, the fog suddenly lifted at this point, and the universe became transparent.

It has remained transparent since. Therefore we can conclude that the microwave background that we see today had its origin about a million years after the beginning. Of course, much of it was originally emitted much earlier than this. However, until the fog lifted, the radiation bounced around so much and underwent so many transformations that all we can really claim to be "seeing" when we look at the background is the universe as it was when it was a million years old.

Some fairly simple calculations indicate that when the universe suddenly became transparent, it was more compressed than it is now by a factor of about a thousand. It doesn't make any difference whether the universe is open or closed; the same result holds in either case. Although an open, infinite universe cannot become "larger" in the ordinary sense of the term, it can still expand, in the sense that matter in it can become more dispersed.

Although the universe was much smaller at the age of a million years than it is now, it had already expanded considerably since the big bang. Consequently, if we are to have any real confidence in the big bang theory, we must find a way to look even farther back in time. After all, although it hardly seems very likely, it is at least conceivable that the universe never was much smaller than it was when it suddenly became transparent. If we did not have any evidence of an earlier history, we could never be entirely sure that the expansion did not begin at this point.

Fortunately, evidence of an earlier history does exist. In fact, it is possible to "see" the universe as it was just three minutes after the big bang. This is the evidence that is derived from studies of the chemical composition of the universe.

The visible matter in the universe is made up primarily of two elements, hydrogen and helium. All of the other elements, including most of those that make up the earth, are present only in small quantities. Of the two major components of the universe, hydrogen is by far the more abundant; the universe is approximately 75 percent hydrogen and 25 percent helium by weight. Approximately the same percentages are observed everywhere, in distant galaxies as well as in our own. It is unlikely that any of the measurements that have been made contain significant errors. It is a fairly simple matter to determine the chemical makeup of a star or a galaxy. One only has to analyze the light that it emits. Furthermore, it is possible to measure the amounts of hydrogen and helium in interstellar space by looking at their radiowave emissions.

The nuclear reactions that take place within the interiors of stars convert hydrogen into helium. The process is similar to those that take place in a hydrogen bomb explosion; hydrogen fusion accounts for most of the energy that is emitted by stars. However, this process proceeds far too slowly to account for the amount of helium that is observed in the universe. The stars have not existed long enough to produce more than a fraction of the helium that is seen.

But the presence of large amounts of helium can be explained by the big bang theory. Calculations indicate that when the universe was about three minutes old, conditions would have been such that helium would have been created. When the universe was less than three minutes old, temperatures were too high to allow the hydrogen fusion reaction to take place; a short time later, the universe was too cool.

This conclusion is confirmed by measurements of the concentration of *deuterium,* an isotope of hydrogen that is present throughout the universe in a concentration of twenty to thirty parts per million. An ordinary hydrogen nucleus is made up of a single proton. A deuterium nucleus, on the other hand, is composed of a proton and a neutron. Deuterium is said to be an isotope of hydrogen because when the nuclei combine with electrons to form atoms, only one electron will attach itself to the nucleus in either case. The negatively charged electron reacts only to the presence of the positively charged proton; it behaves as though the neutron (which has no electrical charge) were not there.

Deuterium cannot be made in stars. A single proton and a single neutron are bound to one another quite loosely; it doesn't take much energy to knock them apart. At the temperatures that exist in stellar interiors, deuterium nuclei break up as soon as they are formed. Thus the deuterium that exists in the universe can only have been made in the primeval fireball. The big bang theory also explains why the concentration of deuterium is so low. The formation of deuterium was one of the steps in the creation of helium. Most of the deuterium disappeared when helium nuclei were formed. The deuterium nuclei that exist today are those that failed to find partners.

If the observation of helium and deuterium concentrations allows us to "see" the universe as it was only three minutes after the beginning, it seems natural to ask whether there might not be ways to look even farther back in time. In principle, this is indeed possible. However, the experimental difficulties are so great that this feat has not yet been accomplished.

Theoretical calculations indicate that approximately one second after the beginning, neutrinos should have been produced in enormous quantities. Since neutrinos rarely interact with matter, there should still be about 100 million of them for every neutron or proton in the universe today. This means that there must be approximately 500 neutrinos for every cubic centimeter of space in the universe, and that about a million trillion neutrinos must be passing through every human body each second.

Unfortunately, no one has ever been able to think of a way that these cosmic neutrinos might be detected. The problem is that the cosmic neutrinos have been losing energy during the 15 billion years that have passed since they were created. At present, their energy should be about a billionth of that possessed by the neutrinos that are by-products of the nuclear reactions that take place in the sun. Since even the relatively energetic solar neutrinos are not easy to detect, observing the cosmic neutrinos is apparently a hopeless task, given the present state of experimental technology.

But if we look a bit farther back in time, other possibilities appear. Calculations indicate that about a thousandth of a second after the beginning, large numbers of quarks, the theoretical constituents of such particles as protons and neutrons, should have been created. Most of these quarks should have combined with one another to produce neutrons, protons, and other particles. However, there is reason to believe that some of them failed to find partners, just as some of the deuterium nuclei that were created three minutes later failed to combine to form helium.

Experiments have been performed in which attempts have been made to detect these relic quarks. However, as I write this, none of them has yet been successful. Nor have physicists been able to detect any of the other *relic particles* that theoretically should still be present today.

Current theories of particle interactions at high energies predict that *magnetic monopoles,* isolated north or south magnetic poles, should have been created in the big bang. But if they were, they were apparently produced in such small numbers that it is not so easy to find them. Nor does it seem possible to detect *gravitons,* theoretical particles associated with the force of gravity, even though the big bang should have produced these particles in enormous numbers.

Naturally the experimental situation could change quite rapidly. The searches for relic quarks and for magnetic monopoles continue, and they may prove to be successful sometime during the next few years. However, until they are, it will be necessary to depend solely upon theory if we want to know what was happening during the first three minutes after the universe came into existence.

I don't propose to discuss the interactions that may have taken place in the early universe in detail. After all, my only purpose in discussing the big bang theory is to attempt to see what implications it might have for our understanding of the nature of time. In any case, there exist a number of excellent books that discuss big bang physics. Some of these are listed in the bibliography that appears at the end of this book. I propose, instead, to leap back to a moment in time 10^{-43} second after the beginning.

At this point in time, the universe was almost unimaginably hot and dense. If current theories are correct, the temperature was approximately 10^{32} K (or $10^{32\circ}$ C; when we speak of temperatures of a hundred million trillion trillion degrees, it is senseless to worry about the 273-degree difference between the two scales), and the density of matter was about 10^{18} (a million trillion) tons per cubic centimeter. Energies were so high that individual particles had energies roughly equal to that of a battleship moving at full speed through the ocean.

At this point a skeptic might object that there is no way that we can really be sure that such conditions existed. If he does, there might be some point to his objections, for the above description depends upon the assumption that Einstein's general theory of relativity remains valid under conditions far more extreme than any under which it has ever been tested. On the other hand, no one has yet discovered anything that would prevent the application of the theory to the early universe at this point in its evolution. So perhaps the best course is to forge ahead and to try to see what the theory predicts. After all, it is the best theory of graviy that we have. Perhaps it might be best not to be too dogmatic about the picture that the theory provides. On the oher hand, there is nothing wrong with attempting to see just what kind of picture this is.

In a previous chapter it was observed that general relativity predicts that singularities should be formed when large, burned-out stars collapse into black holes. Under such conditions, there is no way that singularities can be avoided, if the theory describes such events accurately. During the 1960s the British physicists Roger Penrose and Stephen Hawking proved some theorems that indicated that if general relativity is a correct theory, then singularities become unavoidable when gravity is sufficiently intense. In other words, there seems to be no way that particles of matter collapsing into a black hole can "miss" one another and avoid forming a singularity.

Gravity is just as intense in the early stages of the big bang. Similar theorems show, again if general relativity is correct, that all of the matter and energy in the universe must have been confined in a singularity at time zero. However, these theorems do not really prove that singularities existed. When we approach the quantum region, where general relativity is no longer expected to be valid, they are no longer applicable.

It is expected that general relativity should be valid down to distances of around 10^{-33} centimeters. It is actually possible to be a little more specific than this. There is a quantity, called the *Planck distance,* named after the German physicist Max Planck, who was one of the founders of quantum theory, that is equal to 1.61×10^{-33} centimeters. At distances less than this, a quantum theory of gravity is needed if one is to have any hope of describing physical processes correctly. However, it is expected that the general theory of relativity should break down gradually as this distance is approached. Most likely, it would not suddenly cease to work; its predictions would simply become less and less accurate as distances became smaller. Consequently, no one worries very much about the factor of 1.61; when physicists speak of a point at which general relativity ceases to be applicable, they are only interested in obtaining a figure that is "in the right ballpark."

If one divides the Planck distance by the velocity of light, one obtains a quantity called the *Planck time.* Since the speed of light is 3×10^{10} centimeters per second in metric units, the Planck time is 5.36×10^{-44} seconds. This is the time that it would take light to travel across the Planck distance. Again, if we are only interested in getting the right order of magnitude, the figure can be rounded off to 10^{-43} seconds. It is thought that the general theory of relativity should cease to be valid at times less than 10^{-43} seconds after the beginning of the universe. It is not possible to say anything about events that took place before this time. Perhaps space and time themselves had an entirely different character than they possess now. Indeed, the very concepts of "space" and "time" may cease to have any meaning in this region.

Discussions of the big bang theory often make use of such terms as "the beginning," "the creation of the universe," and so on. The use of such terminology does not imply that the big bang theory states that the universe was created at a given instant in time. It says nothing of the sort. All that the theory can really tell us is that at the Planck time, the universe was in an extremely hot, dense state. The Planck time is a point beyond which we cannot see; it is the point at which all of known physics breaks down.

Naturally it is possible to make conjectures about events that took place before the Planck time. But when one does this, one is engaging in philosophy, or possibly theology; one is no longer doing physics. However, if we keep this reservation in mind, it is possible to enumerate the various different possibilities. There seem to be three of these: (1) The universe was created at a given instant in time. (2) The universe existed in some unknown form before the big bang. (3) Space and time were themselves created in the big bang.

Many physicists find the third possibility to be particularly appealing. It avoids all of the philosophical problems associated with the contemplation of an infinite past, as well as those associated with creation at a particular point in time (for example: "Why was the universe created at that instant and not a billion years sooner?"). And, of course, this third possibility is scientifically plausible as well. After all, the big bang was not an event that caused the universe to expand explosively outward into space. On the contrary, the big bang was an event that happened everywhere. Space itself expanded with the universe.

In other words, the big bang may have been the beginning of time. In this case, the question "What happened before the big bang?" would be meaningless; there would be no such thing as "before."

Although astronomers and cosmologists generally accept the big bang theory in its broad outlines, they realize that the theory does contain difficulties; there are a number of phenomena that it leaves unexplained. For example, it does not tell us why the average curvature of space-time should be so close to zero.

Astronomers have not been able to tell whether we live in a closed, positively curved universe or in an open, negatively curved one. All that they can tell is that the density of mass in the universe is somewhere between one-tenth of the critical value and ten times the critical value. For all practical purposes, the universe is teetering on the borderline.

At first glance, an uncertainty that is equal to a factor of ten in either direction does not seem to be so small a quantity. However, calculations indicate that in an earlier era, the universe must have been much more delicately balanced. In order to be that close to the borderline today, the universe would have had to be expanding, at the Planck time, at a rate equal to a critical value to an accuracy of one part in 10^{60}. If it had been expanding just a little too slowly, gravity would have quickly halted the expansion, and the universe would have recollapsed in what has been called, somewhat facetiously, a "big burp." If it had been expanding a little too rapidly, matter would have become dispersed far too rapidly for galaxies to form. Since rate of expansion and mass density are related, the fine-tuning must have been very precise in order to create the density that is observed today.

When physicists find that a quantity has a very precise value, they generally seek a reason for this. The failure of the big bang theory to explain why the average curvature of the universe is so close to zero must be considered a defect. Of course, one could say, "Maybe it was simply a coincidence." But that would be a rather lame attempt at finding an explanation. After all, it is not likely that a simple coincidence would turn out to be so precise.

Another problem with the big bang theory is that it does not explain why matter in the universe should be so evenly distributed. Admittedly, the distribution doesn't look very uniform to the naked eye. When one looks up into the sky, one does not see a uniform haze; matter seems to be concentrated into stars. However, when one looks at the universe as it appears on a very large scale, uniformity does begin to appear. Galaxies seem to be rather uniformly distributed once one progresses to distances of a billion light-years or more. Galaxies begin to take on an appearance that is not unlike that of grains of sand on a beach. Although the grains might appear to be irregularly distributed boulders to an ant, a beach can appear to be fairly flat and uniform to a human who gazes out over distances of hundreds of yards. And of course the microwave background is part of the universe too. It is even more uniform than the galaxies. As I noted previously, it varies by about three parts in ten thousand.

The uniformity of the universe does not seem to be much of a problem until one considers the fact that according to the standard big bang theory, there are large portions of the universe that have never been in contact with one another. If the universe is 15 billion years old, then regions of the universe that are more than 15 billion light-years apart cannot influence one another. It takes light 15 billion years to travel that distance, and causal influences cannot travel faster than light. Furthermore, these regions could not have been in contact with one another in the past. During earlier stages of the evolution of the universe, the relevant distances were even smaller. For example, when the universe was one second old, regions that were more than one light-second (186,000 miles) away from one another were not in causal contact.

And yet these distant regions must have been in contact to produce the uniformity that is observed. As far as we can tell, the universe—at least when it is viewed on a large scale—is expanding at the same rate in every direction; wherever we look, matter and radiation have about the same density. There must have been some influencing factor that caused far-flung regions to behave in so coordinated a manner.

Finally, the standard version of the big bang theory does not explain why there should be galaxies. When detailed calculations are made, one obtains the result that if galaxies were to form, the universe must have been "lumpy" during the early stages of the expansion. If it was not, particles of matter would have expanded away from one another before gravitation had a chance to cause them to coalesce into galaxies. But the theory does not seem to be capable of explaining where these lumps came from.

In 1980, physicist Alan H. Guth of the Massachusetts Institute of Technology proposed a new hypothesis, an elaboration of the big bang theory, that would remove some of these difficulties. Guth's *inflationay universe* theory is a speculative one, and it is based on theories that are themselves speculative. Nevertheless, his idea is quite a plausible one. Although there is, as yet, no experimental evidence to support it, it does have a certain intuitive appeal.

In order to see what Guth's theory is all about, it will be necessary to embark on a digression and to say something about the forces that exist in nature. Physicists have discovered four of these: gravity, the electromagnetic force, the strong nuclear force, and the weak nuclear force. The electromagnetic force is responsible for all known electrical and magnetic phenomena, and for the existence of electromagnetic radiation as well. The strong nuclear force is the one that binds protons and neutrons together in atomic nuclei, and the weak nuclear force is responsible for certain kinds of nuclear processes, such as beta decay.

Physicists would like to discover a single theory that would explain all four forces. If they are ever successful, they will have a single theory which explains the behavior of matter in an elegant and parsimonious way. Furthermore, a theory that unified all four forces would almost certainly predict important but as-yet-undiscovered phenomena. At least this is what has happened when theories that unified the various forces of nature have been discovered in the past. For example, when Maxwell found a unification of electric and magnetic forces in the nineteenth century, the resulting theory not only showed that light was a form of electromagnetic radiation, it also predicted the existence of radio waves.

The electromagnetic and weak forces were joined in 1967 when the American physicist Steven Weinberg and the Pakistani physicist Abdus Salam independently discovered a theoretical unification of the two interactions. Since that time, experimental confirmation of the Weinberg-Salam theory has been obtained; among other things, the theory predicted the existence of previously unknown particles that have since been observed in the laboratory.

Theoretical physicists are currently seeking to go a step further and to find a unification of the electromagnetic, weak, and strong forces (there are some especially difficult problems associated with gravity; it is likely that it will be the last to be included in a unified theoretical scheme). A number of theories that attempt to carry out this unification already exist. However, these *grand unified theories,* or GUTs, as they are often called, are somewhat speculative. No one knows which, if any, of them is likely to be correct.

Even though there are a number of GUTs, they bear a family resemblance to one another, and it is possible to make certain kinds of theoretical predictions without worrying about which particular one is most likely to be correct. For example, the GUTs predict that the proton, which was previously thought to be one of the few perfectly stable particles, should decay into certain other particles after 10^{30} to 10^{32} years. Experimental physicists are currently attempting to verify this prediction in the laboratory. As I write this, they have not yet been successful.

A successful grand unified theory would be particularly useful in explaining events that took place early during the history of the universe. GUTs theories, after all, are designed to explain the properties of matter at high energies, and the energies of the particles that existed during the first thousandth of a second after the creation of the universe are far higher than any that could conceivably be produced in the laboratory.

But Guth was less interested in the implications of the GUTs for the behavior of individual particles than he was in the behavior of the early universe itself. When he looked at this question in detail, he found that the GUTs implied that the universe must have gone through a phase of particularly rapid expansion when it was approximately 10^{-35} seconds old. According to Guth, there was a period of *inflationary expansion* that lasted for about 10^{-32} seconds and that caused the universe to increase in size by a factor of 10^{25} or more.

According to Guth, the energy generated during this phase of rapid expansion may be the source of most of the matter and energy in the universe today. According to Einstein's famous equation $E = mc^2$, matter and energy are equivalent; consequently, the production of energy will lead to the creation of matter. According to Guth, all that is needed to explain the present universe is the assumption that there once existed a hot, strongly curved bubble of space-time that contained perhaps twenty pounds of matter; all the rest could have been created from the energy produced by the inflation. For that matter, Guth adds, the inflationary expansion could be the source of *all* matter and energy. It may not even be necessary to assume that those twenty pounds were present when the inflation began. "The universe," Guth says, "could be the ultimate free lunch."

Is Guth suggesting that the universe may have been created out of nothing? The answer to that question must be yes and no. It is possible that the universe grew from a piece of space-time that was devoid of matter. But that isn't quite "nothing." According to the general theory of relativity, the curvature of space-time can itself be a source of energy, and changes in the curvature should cause the creation of matter. In Einstein's theory, space-time can have quasi-material properties.

The inflationary universe theory sounds quite fantastic. But this should not be regarded as a criticism. Einstein's theories seemed fantastic too, when they were first proposed, and so did quantum mechanics. Yet all of these theories were subsequently confirmed. Sometimes it is only the fantastic theories that have any chance of being true; sometimes it is necessary to discover new ways of thinking about the processes that take place in nature if one is to have any hope of understanding what is going on. These new ways of thinking must necessarily seem bizarre at first. Physicists generally recognize the fact that this is the case. For example, the Danish physicist Niels Bohr, one of the founders of quantum mechanics, once criticized a new theory on the grounds that it was not "crazy enough."

Although there is not yet a shred of evidence to support Guth's theory, the inflationary universe idea does seem capable of solving some of the problems that seem so difficult when one attempts to rely on the standard big bang theory. The inflationary universe theory solves the *flatness problem,* the problem of why the average curvature of space should be so close to zero, quite nicely. Calculations indicate that at the end of the inflationary period, the universe must have been expanding at precisely the right velocity; the dynamics of the inflation bring this about. The fact that the universe is so close to the borderline between negative and positive overall curvature is an automatic consequence of the inflationary expansion.

Guth's theory also explains the evenness of the expansion and of the distribution of radiation and matter. If the universe expanded as much as Guth's theory indicates it did during the inflationary period, then it must have been much smaller before the beginning of that period than cosmologists had previously supposed. Before the inflation began, different portions of the universe would have been in causal contact with one another, and there would have been an opportunity for irregularities in the energy content of the universe and in the expansion rate to average out.

The inflationary universe theory is somewhat less successful in explaining the existence of galaxies. The theory does imply that there would have been fluctuations during the inflationary period that could have led to regions of higher-than-average matter density later on. But it predicts conglomerations of matter that are much larger than galaxies. However, this problem may eventually be solved. In any case, the inflationary universe theory seems better able to explain the existence of galaxies than the standard big bang theory.

Finally, Guth's theory may explain why physicists have not yet been able to confirm the existence of magnetic monopoles. According to the standard theory, monopoles should be ubiquitous. But according to the inflationary universe theory, relatively few of them were produced.

All in all, the inflationary universe theory is quite appealing. It is difficult to believe that a theory that explains so much does not contain at least an element of truth. Of course, until one or another of the grand unified theories is confirmed, the inflationary universe theory will lack a solid foundation. Nevertheless, the theory must be taken seriously. Even if it turns out to be incorrect in some of its details, it may still point the way toward an understanding of some of the events that took place shortly after the big bang.

The inflationary universe theory also makes it possible to engage in speculation about such matters as the beginning of time. In order to undertake such speculation, it is necessary to imagine events that may have been taking place before the Planck time. However, if one has an idea as to what may have been happening when the universe was 10^{-43} second old, it is possible to make guesses as to what might have been going on earlier.

When we attempt to look back beyond the Planck time, we have no theory to guide us; we are looking at a region of time where conditions were so extreme that all known physics ceases to be applicable. However, such speculation is not completely fantasy. One cannot imagine anything that one wants; whatever we come up with must be consistent with current ideas about the events that took place after the Planck time.

If the inflationary universe theory is correct, then at a time of 10^{-43} seconds, the universe consisted of a tiny bubble of space-time that contained little or no matter. It is possible that before the Planck time, there existed a kind of primordial chaos. Of course such a chaos, if there really was one, did not exist in space or in time; neither had yet been created.

It is possible to imagine that small bubbles of space-time may have been created out of this chaos. Some of them were contracting, some were expanding; most of them disappeared almost as soon as they were formed. But every now and then (we should probably think of "now" and "then" as metaphors rather than as words with definite meaning; time may not have existed yet) some of the bubbles remained in existence for a period longer than the Planck time. Some of these soon disappeared, but others continued to expand slowly, until they were caught up in the inflationary expansion that Guth's theory says would begin at about 10^{-35} seconds. One such bubble became the universe we live in today.

Such a scenario implies, of course, that there might be a large number of other universes, perhaps even an infinity of them. It would naturally not be possible to speak of "where" these universes are. After all, space-time might exist only within the universes. To apply such terms as "where" or "when" to them would then be meaningless.

At first glance, it all sounds rather bizarre. But perhaps it is really not more difficult to believe in many universes than it is to think that ours was the only one ever created. After all, if one universe can come into existence in this manner, there is apparently no reason why the same events could not take place on countless different occasions.

Of course, this is not the only possibility. It is also possible to imagine that our universe has existed, in one form or another, for an infinite period of time, that it may not have been spontaneously formed from a space-time bubble. I propose to examine this idea in more detail in the next chapter. For the moment, I am only concerned with speculating about the possibility that time (or more precisely, space-time) may have had a beginning, and that it may have an end.

If the universe is closed, there may be not only a beginning but also an end of time. A closed universe must eventually collapse upon itself. Since astronomers' attempts to measure the precise rate at which the expansion of the universe is slowing down have so far not been very successful, it is not possible to say exactly when this collapse would take place. However, since it appears that the universe will continue expanding for some time yet, a closed universe would probably not begin to contract for at least another 40 or 50 billion years.

At first, the contradiction would be a very slow process. But then, as galaxies moved closer and closer together, gravitational attraction would become stronger, and the collapse would proceed at an ever-increasing rate. If nothing happened to halt this process, conditions would eventually become similar to those that existed shortly after the big bang. As the big crunch continued, matter in the universe would become more and more compressed. The collapse of a closed universe is not unlike the collapse of a dead star into a black hole. Eventually, all of the matter in the universe would be crushed out of existence in a singularity, or into a region of space-time with dimensions less than the Planck distance. If the universe did indeed begin as a bubble of space-time in a primordial chaos, the big crunch might cause it to return to the chaos from which it came.

There is a possibility that there might be unknown processes that would cause the universe to "bounce" and to re-explode in a new big bang. If this does not happen, then the big crunch would—in all probability—represent the end of time. If all of the matter and energy in the universe were crushed into nothing, it is likely that space-time itself would cease to exist.

If, on the other hand, the universe is open, the present expansion will continue indefinitely. Galaxies will continue to recede from one another; one by one, the stars will burn out. New stars will be created, but the supplies of interstellar gas from which stars are formed will eventually be used up. When the universe is about 100 trillion years old—about ten thousand times as old as it is now—the sky will become completely dark.

As the stars die, some of their planets will be vaporized when stars go through the red giant stages of their evolution. The remaining planets will be knocked out of their star systems by interstellar collisions after about 10^{15} to 10^{17} years. Such collisions do not happen very frequently, but they will inevitably take place if the universe lasts for long enough a time.

Random collisions will also cause the galaxies to lose some of their stars by *galactic evaporation*. This process is similar to the evaporation of molecules from a liquid. In each case, collisions cause some of the stars or molecules to gain enough energy to escape from the attractive forces exerted by the rest. As these stars—perhaps as many as 90 percent in a typical galaxy—fly off into intergalactic space, the remaining stars will be drawn into the galactic cores. Even if supermassive black holes do not exist in the centers of galaxies at present, they will certainly be formed at this stage of the evolution of the universe.

After about 10^{18} (a million trillion) years, nothing will be left of the universe but the remains of collapsed galaxies and the various objects that will now be scattered through space: dead stars, stellar-sized black holes, planets, and smaller objects such as comets.

It is possible to continue the scenario even further. But

before I do, it might be well to digress a bit and to make a few remarks about the probable accuracy of predictions about events that might take place far in the future. These predictions are necessarily based on the assumption that the known laws of physics can be applied to events that would take place over immensely long periods of time. Naturally one cannot be perfectly sure that this is legitimate. The problems associated with speaking of events that will take place 10^{18} years from now introduce a great deal of uncertainty.

The problem is that we cannot be sure that the laws of physics will remain the same over such long periods of time. For example, it is conceivable that the strength of the force of gravity could change. It has been determined that if gravity is gradually becoming stronger or weaker, this must be a very slow process. Experiments have been performed showing that if the force of gravity is changing, the change must be less than twenty or thirty parts per trillion each year. But suppose that gravity was becoming stronger or weaker at a rate too small to be measured by presently available methods. The change might be too small to have a significant effect on the structure of the universe a billion years from now, or even in 10 or 100 billion years. However, periods of trillions or millions of trillions of years are an entirely different matter. If we are to make any predictions at all, we must assume that the force of gravity remains *precisely* the same, and we must assume that the other forces in nature are not altered either. No one knows whether or not this assumption is really valid.

When one speaks of very long periods of time, it is also necessary to assume that there are no unknown processes that might affect one's results. But there is no way of being sure that this assumption is valid either. For example, the above discussion depends upon the assumption that nothing will happen to affect the stability of ordinary matter during the next 10^{18} years. But this might not be true. For example, it has been suggested that magnetic monopoles could catalyze the decay of nuclear particles over long periods of time. If monopoles really do exist, and if there are more of them than the inflationary universe theory predicts, these decays could cause matter to disintegrate long before stars began to lose their planets or galaxies to lose their stars. And, of course, there could be any number of as-yet-undiscovered processes taking place in our universe that could also affect the future stability of matter.

However, if we keep these reservations in mind, it is possible to look even farther into the future of an open universe. After all, there is nothing wrong with attempting to see what currently accepted theories predict. The picture they paint may not turn out to be perfectly accurate. However, the only alternative would be to throw up our hands in dismay and refrain from constructing any scenario at all.

If the predictions of the GUTs are correct, and the proton decays in 10^{30} to 10^{32} years, the remaining planets, and the stars that have not collapsed into black holes, will disappear at about this time. As the protons decay, stars and planets will disintegrate into subnuclear particles. Many of these particles will decay in turn, until there is nothing left of the universe but black holes, electrons, positrons, neutrinos, and antineutrinos, and a bit of radiation of extremely weak intensity.

The black holes will not last forever. When one applies quantum mechanics to black holes, one obtains the result that the black holes should eventually evaporate into streams of radiation and particles. The amount of time required before this will happen is extremely large. Only after something like 100^{100} years have passed will the largest black holes, the supermassive ones in the cores of galaxies, disappear. But if the predictions of quantum mechanics are correct—and if it is possible to make any kind of prediction about events that will take place after a time that is 10^{90} times as long as the universe has so far existed—they will inevitably encounter this fate.

Eventually, there will be nothing left but electrons, positrons, neutrinos, antineutrinos, and radiation. But, after another 10^{70} years or so, the positrons and electrons will annihilate one another, and bursts of gamma rays will appear in their place. As this happens, the universe will, of course, continue its expansion. Thus the gamma rays will experience a red shift, and their energy will gradually decrease as they are progressively transformed into X rays, ultraviolet radiation, visible light, infrared radiation, and, finally, radio waves. In the end, there will be nothing left in the universe but the neutrinos and antineutrinos and perhaps a few electrons and positrons that have escaped annihilation. As the expansion goes on, the average distance between these particles will continue to increase until, for all practical purposes, there will be nothing left of the universe but empty space.

Thus it appears that time will eventually come to an end in an open universe too. As the universe becomes emptier and emptier, as it evaporates into nothing, events will cease to take place. Without events to mark its passing, time cannot be measured, or even defined. Perhaps one could relate the passage of time to the continuing expansion of the now-almost-empty universe. It is clear, however, that this kind of time—if indeed one can call it time—bears little relation to the variety that we measure on earth.

CHAPTER 12

What Is Time?

ISAAC NEWTON THOUGHT HE knew what time was. "Absolute, true and mathematical time," he wrote at the beginning of his *Principia,* "of itself, and from its own nature, flows equably without relation to anything external."

Today we know that Newton was mistaken in several different respects. Time is not absolute; it is relative. As the special theory of relativity shows, time measurements depend upon the state of motion of the observer. Time is not a substance that "flows equably without relation to anything external." According to the general theory of relativity, the presence of matter creates gravitational fields that cause time dilation.

Finally, if time does "flow," this flow cannot be measured by experiments in the laboratory. The movement of the "now" from the present to the future seems to be a subjective phenomenon. In physics, there is no concept of a "present moment"; the mathematical equations associated with physical laws and theories deal only with the intervals between instants of time. There is no standard by which a flow of time can be measured. At best, one can only say that time moves onward at the rate of one second per second, which is about as meaningful as defining the word "cat" by saying, "A cat is a cat." Nor can any meaning be attached to the statement that time "flows equably." If the flow of time is not uniform, how can one measure its irregularities?

Physicists do speak of the arrows of time. However, this concept has nothing to do with a flow. When we speak of "time's arrows," we mean only that the world has a different appearance in one direction of time than it does in the other. Perhaps it would not be inappropriate to illustrate this idea with an analogy. Suppose that a woman is standing on a beach with her back to the ocean. If she looks straight ahead at the land, and then turns around and looks at the ocean, things will have a different appearance. There will be an asymmetry along one of the directions of space. Yet the existence of this asymmetry does not imply movement in one direction or the other. The observer can see it while she is standing still.

If Newton's definition is inadequate, how would one define time? One cannot say that it is something that defines the temporal order of events. Even if one ignores the circularity implied by the word "temporal," there is the problem that the time order of events can depend upon the state of motion of an observer. Special relativity tells us that if event A precedes event B in the reference system of one observer, this order may be reversed in the reference system of another.

The best solution might be to view time from the standpoint of general relativity and to look at it as an element in the four-dimensional geometry of space-time. But even then there are unanswered questions. No one knows whether the concept of "time" even has any meaning when one enters the Planck region. When one looks at time on a scale of 10^{-43} second or less, "time" may cease to exist. Alternatively, time on that scale could conceivably be made up of small particles, sometimes called *chronons.* It may be that the time dimension does not have a smooth, continuous structure. It is possible that time could be made up of particles that are so tiny that we have not yet been able to detect them. In other words, time could be made up of discrete moments, with nothing in between them.

The general theory of relativity seems to provide the most complete description of time. Yet the theory does not tell us why there should be arrows of time, or what the arrows have to do with one another. The relationship between general relativity and the arrow of time in thermodynamics is especially puzzling. Since physicists are not quite sure how one would go about calculating the entropy of a gravitational field, or even whether the concept of entropy can be applied to gravity, it is not easy to see what the dynamics of the universe have to do with the direction of time.

The existence of this difficulty has led to some bizarre speculations. For example, it is possible to ask whether the arrow of time might not be reversed in a contracting universe. If there is some connection between the increase of entropy and the universe's expansion, then it would seem that it might be reasonable to conclude that entropy would decrease in a contracting universe. Then, if entropy is still used to define the arrow of time, one would have to conclude that time would start to go backward when the phase of contraction began.

In such a contracting universe, light would not flow out of stars, it would stream into them. Rivers would run uphill, and rain would rise from the earth to the sky. But if intelligent beings still existed at this point in the evolution of the universe, they would notice nothing unusual about any of this. Since the processes taking place in their brains would also be reversed, they would "think backward" and view phenomena exactly as we do. Furthermore, they would see the contraction of the universe as an expansion; they would look back to the big crunch and view it as the beginning of the universe, not the end.

If time did reverse itself at the moment of maximum expansion, the distinction between a contracting and an expanding universe would be purely arbitrary. There would be no way of telling in which half of the universe one was living. There would be no way to distinguish between forward and backward time; it would all depend upon one's point of view. Consequently, time in such a universe would not have an end; instead, it would have two beginnings.

It is probably safe to say that most physicists do not take this possibility very seriously. Although it is possible to conceive of a universe with two big bangs and no big crunch, the theory does contain difficulties that are not easily removed. For example, consider a time a few million years after the expansion of the universe had reversed. Since time's arrow would have reversed, light would now be flowing back into the stars. But light emitted by stars in galaxies hundreds of millions or billions of light-years away would still be arriving. To an observer in the time-reversed universe, it would appear that radiation was traveling both into the future and into the past. This is a rather paradoxical conclusion, especially when one considers the fact that an observer living a few million years before the expansion came to a halt would view phenomena in a normal manner.

Although certain theoretical problems exist, there really does not seem to be any way to demonstrate that time could not reverse itself at the moment of maximum expansion. I suspect that our inability to exclude this possibility may be a reflection of our lack of knowledge about the fundamental nature of time. If physicists knew more about the connection between gravitation and entropy, they might be able to prove that such a paradoxical kind of universe was impossible.

There are other problems associated with attempts to apply the second law of thermodynamics to the universe as a whole. Presumably the universe began in a very chaotic state. A chaotic state is, by definition, a state of high entropy (when we speak of "chaos," we mean that there is a great deal of disorder). On the other hand, numerous kinds of structure have appeared since the universe began. For example, stars and galaxies have formed. The creation of this structure, and the fact that stars gain entropy as they burn their nuclear fuel, would seem to imply that the universe is far from a state of maximum entropy now. But how can this be, if entropy was so high at the beginning? Doesn't the second law of thermodynamics tell us that entropy always increases with time?

There have been attempts to come to terms with some of these problems. For example, in 1983 the British physicist Paul Davies suggested that the inflationary universe theory might explain why the universe is in a relatively low entropy state. According to Davies, the entropy of the universe may indeed have been near a maximum at the Planck time. However, Davies says, the inflationary expansion that took place shortly afterward could have created an "entropy gap." He suggests that although entropy increased during the phase of inflationary expansion, the amount of possible entropy increased even more. The increase of entropy in the inflationary universe was analogous to pouring water into a container that grew larger at a faster rate than the water could be poured. Thus, at the end of the inflationary period, entropy was higher than it had been previously, but it was not near a maximum anymore. Davies goes on to suggest that it was the inflation of the universe that fixed the arrow of time. He feels that it was the inflation that provided room for entropy to increase.

Other physicists are not so sure. Many of them think that the universe started out in a low entropy state, and various hypotheses have been constructed that attempt to describe what this low entropy state may have been like. But it is not likely that the controversy will be settled anytime soon. Until the puzzle of gravitational entropy is solved, no one will really know which of the various hypotheses has the best chance of being true.

Sometimes it seems that scientific investigations into the nature of time reveal more about what time is not than they do about what time is. In fact, physics reveals that the very concept of time contains apparent contradictions that are not easy to resolve. The special theory of relativity tells us that time is not a substance that "flows equably" throughout the universe. But general relativity seems to imply, at least when it is applied to the early universe, that space-time has a quasi-material character. The existence of the arrows of time seems to imply that time should have a fixed direction. And yet it is possible to view the positron as an electron moving backward in time. The second law of thermodynamics seems to say that the arrow of time is a macroscopic, statistical phenomenon. And yet the time asymmetry exhibited by a single unimportant particle, the K meson, tells us that there is sometimes an arrow of time on the subatomic level also. Finally, so little is known about the origin of the arrows of time that it is impossible to demonstrate that time could not reverse its direction in a contracting universe.

It is not even possible to tell whether time has a beginning and an end, or whether our universe has existed for an infinite length of time. In the previous chapter, I explored the ways in which time might have begun. But as we saw, the theories that describe the origin of the universe are quite speculative. It is perfectly possible that time extends into the infinite past, and into the infinite future as well. The idea that space and time were created in the big bang is philosophically appealing. But it does not necessarily follow that it is true.

Since the 1930s, physicists and cosmologists have been speculating about the possibility that the universe might go through an endless series of expansions and contractions. Such *oscillating universe* theories are based on the assumption that the universe is closed, and that it is not destroyed in a big crunch. It is assumed that the universe somehow "bounces" and explodes outward in a new big bang. No one really knows how such a bounce might take place. However, there is nothing wrong with making the assumption that it might be possible. After all, when one encounters a situation where there is a great deal of uncertainty as to what might have been going on, it is perfectly legitimate to speculate about the various different possibilities. It would be difficult to discover the truth about the universe if we refused to consider anything that might be true.

According to Alan Guth, the inflationary universe theory implies that the universe cannot bounce back out of a big crunch. However, this is a controversial point. The behavior of the universe in such a situation depends on the particular assumptions that one makes, and no one is sure which assumptions are the correct ones. Furthermore, it is possible to construct theories of gravity that predict that the universe would bounce if it contracted to a certain minimum radius. Such a theory, called *nonsymmetric gravitational theory,* has been constructed by physicist John William Moffatt of the University of Toronto, for example. Although Moffatt's theory is somewhat more complicated than Einstein's, there is really no available experimental evidence that would rule it out. Thus it seems that the possibility of a bouncing universe cannot be eliminated.

A bounce would be consistent with Einstein's general theory of relativity, for that matter, if it was an event that took place after the universe had contracted to dimensions less than the Planck distance. Since general relativity cannot tell us what happens in this region, it has nothing to say on the question of whether quantum effects could cause such a bounce to take place. Perhaps they could, perhaps they couldn't. However, since our primary concern is that of finding out what the consequences of a bounce might be, it might be best to forge on, without worrying overmuch about the details of the process.

At first, the idea of a bouncing universe seems somewhat appealing. If the universe does bounce, we can look forward to a universe that will go on forever, in which there will be no end of time, and in which conditions may continue to be such that they are hospitable to life. However, when one considers the fact that entropy would presumably increase from cycle to cycle in an oscillating universe, problems begin to appear. Detailed calculations indicate that an oscillating universe with increasing entropy would not experience cycles of constant length. With each cycle, the time between the beginning of the expansion and the final collapse would grow longer. The behavior of such a universe would be the opposite of that which we observe in a bouncing ball. Where the ball never reaches the same height that it attained on the first bounce, such an oscillating universe would bounce "higher" each time.

It has been calculated that if we live in such a universe, there have been at most a hundred cycles. Since a hundred cycles necessarily have a finite length, an oscillating universe would have to have an origin in time. Since an origin in time is one of the things that the oscillating universe theory was constructed to avoid, this is hardly very satisfactory. Nor would such a bouncing universe remain hospitable to life. As the entropy increased, the creation of life in future cycles would become progressively more unlikely. Black holes would presumably accumulate in an oscillating universe also, if one makes the reasonable assumption that they survived the bounce. Eventually they would become numerous indeed, so numerous that there would not be enough matter that was not in black holes to form stars and galaxies.

There is no evidence that we do not live in a bouncing universe of this sort. However, it does appear that if we want to speculate on the question of whether time might not extend into the infinite past, we may have to modify some of our assumptions, or pursue a different path.

It is possible to do this. For example, physicist John Wheeler suggests that the very laws of physics might change every time the universe collapses and reexpands. In Wheeler's *superspace* theory, the universe is recreated in a different form in each cycle. The reprocessing of natural laws ensures that entropy and black holes do not accumulate for the simple reason that the laws that govern entropy increase and black-hole formation would change every time the universe went through a big crunch.

Wheeler's theory is quite speculative. But then so are the theories that assume that the universe can go through only one, or at most a limited number, of cycles. Since there is no evidence that would allow us to decide between them, it appears that the only conclusion that one can reach is that there is no way of knowing when time began. We may live in an open universe that was created 15 billion years ago, and that has an infinite future. We may live in a closed universe that began at a given moment in time, and that will end at another moment in time. Or we may live in a cyclical universe of some sort.

Even if our universe was created out of nothing, and even if it is open, cycles of a sort might exist. After all, universes could be created on a very large number, perhaps an infinite number, of different occasions. Of course, such universes would neither succeed one another in time nor coexist in a temporal sense. If time exists only within universes, it would be impossible to order them along a time dimension. However, if there are many universes, we would have a situation that would not be so very different from that in which a series of cycles followed one another in time.

The argument that because something happens once it may happen countless times is not exactly an ironclad one. Nevertheless, it does seem to imply that the idea of other universes should at least be considered. Of course, when one speaks of a "universe" in this sense, one can no longer define the word as "everything that exists." One would have to redefine the term to mean "a self-contained region of space-time." But there is no reason why that could not be done. As Lewis Carroll's Humpty Dumpty pointed out, one can define a word to mean whatever one wants.

It might be noted that a similar argument is used by those who would urge the plausibility of the existence of extraterrestrial life. Since there is no evidence either that life exists elsewhere in the universe or that it does not, it is often pointed out that the fact that life evolved on earth at least makes it seem reasonable that it could arise in many other places also. When conclusive arguments do not exist, one generally has to resort to plausible ones.

Perhaps the universe oscillates, perhaps it does not. But even if it doesn't, there could still be many universes, some of them possessing planets very much like the earth. If there are, we would have something bearing a certain resemblance to the cyclical sort of universe. We could not say that time was cyclical in the strict sense of the term. One cycle of events would not follow another. Nevertheless, time would have a character that was closer to the cyclical time of the ancients than to the linear time of Western civilization. Similar cycles could be repeated without end.

When I say this, I don't mean to imply that the ancients, in their wisdom, were aware of a truth that we are only now discovering. That would be nonsense. It is clear that the ancient idea of cyclical time was derived from the observation of cyclical processes in nature, such as the apparent revolution of the stars around the earth, the phases of the moon, and the alternations of the seasons. Modern conceptions of cyclical time, on the other hand, are based on sophisticated theories about the nature of the universe.

Perhaps it should be emphasized that we don't know that time is cyclical. There is still the possibility that time—and the universe—is something that happens only once. But maybe it is not surprising that we can't be sure whether the cyclical or the linear conception of time is the more valid. It appears that the more one examines the concept of time, the more unanswered questions there are.

I don't want to close by giving the impression that I think that contemplating the mysteries of time should evoke a feeling of awe in us. On the contrary, finding out what the relevant questions are often makes a subject seem less mysterious, even when those questions do not yet have answers. Though science has not yet probed all the depths of the subject of time, it at least knows what we should be asking about the subject. Knowing what to ask is often the most significant step on the road to understanding.

Bibliography

Aaronson, Marc; Huchra, John; and Mould, Jeremy. "The Infrared Luminosity/H_1 Velocity-Width Relation and Its Application to the Distance Scale." *Astrophysical Journal*, Vol. 229 (1979), pp. 1–13.

———. "A Distance Scale from the Infrared Magnitude/H_1 Velocity-Width Relation. I. The Calibration." *Astrophysical Journal*, Vol. 237 (1980), pp. 655–65.

Albritton, Claude C., Jr. *The Abyss of Time*. San Francisco: Freeman, Cooper & Co., 1980.

Angrist, Stanley W., and Heplor, Loren G. *Order and Chaos*. New York: Basic Books, 1967.

Aristotle. *Physics*. Lincoln: University of Nebraska Press, 1961.

———. *Problemata*. Oxford: Oxford University Press, 1927.

Augustine, Saint. *The City of God*. New York: Modern Library, 1950.

———. *The Confessions of St. Augustine*. Garden City, N.Y.: Image Books, 1960.

Barrow, John D., and Silk, Joseph. "From Quark to Cosmos." *Nature*, Vol. 308 (1984), pp. 13–14.

Bludman, S. A. "Thermodynamics and the End of a Closed Universe." *Nature*, Vol. 308 (1984), pp. 319–22.

Boas, Maris. *The Scientific Renaissance*. New York: Harper & Row, 1966.

Brandon, S. G. F. *History, Time and Deity*. Manchester, England: Manchester University Press, 1965.

Büchel, W. "Entropy and Information in the Universe." *Nature,* Vol. 213 (1967), pp. 319–20.

Buckley, Jerome Hamilton. *The Triumph of Time.* Cambridge, Mass.: Harvard University Press, 1966.

Burchfield, Joe D. *Lord Kelvin and the Age of the Earth.* New York: Science History Publications, 1975.

Bury, J. B. *The Idea of Progress.* New York: Dover, 1955.

Butterfield, Herbert. *The Origins of Modern Science.* New York: Macmillan, 1962.

Callahan, John Francis. *Four Views of Time in Ancient Philosophy.* Westport, Conn.: Greenwood Press, 1979.

Campbell, Joseph, ed. *Man and Time.* Princeton: Princeton University Press, 1983.

Canuto, V., and Hsieh, S. H. "Case for an Open Universe." *Physical Review Letters,* Vol. 44 (1980), pp. 695–98.

Cipolla, Carlo M. *Clocks and Culture.* New York: Walker, 1967.

Coe, Lee. "The Nature of Time." *American Journal of Physics,* Vol. 37 (1969), pp. 810–15.

Cornford, Francis Macdonald. *Plato's Cosmology.* London: Routledge & Kegan Paul, 1937.

Cumont, Franz. *Astrology and Religion Among the Greeks and Romans.* New York: Dover, 1960.

d'Abro, A. *The Rise of the New Physics.* 2 vols. New York: Dover, 1952.

Dante. *The Divine Comedy.* New York: Rinehart, 1954.

Darwin, Charles. *The Autobiography of Charles Darwin.* New York: Norton, 1969.

Davies, Paul. *The Edge of Infinity.* New York: Simon & Schuster, 1981.

———. *God and the New Physics.* New York: Simon & Schuster, 1983.

———. "Inflation and Time Asymmetry in the Universe." *Nature,* Vol. 301 (1983), pp. 398–400.

———. "The Inflationary Universe." *The Sciences,* Vol. 23, No. 2 (March/April, 1983), pp. 32–37.

———. *The Physics of Time Asymmetry.* Berkeley: University of California Press, 1977.

———. *Space and Time in the Modern Universe.* Cambridge, England: Cambridge University Press, 1977.

Davis, Marc, et al. "On the Virgo Supercluster and the Mean Mass Density of the Universe." *Astrophysical Journal Letters,* Vol. 238 (1980), pp. L113–16.

de Santillana, Giorgio, and von Dechend, Hertha. *Hamlet's Mill.* Boston: Godine, 1977.

de Vaucouleurs, G. "The Cosmological Distance Scale: A Comparison of Two Approaches to the Hubble Constant." *Annals of the New York Academy of Sciences,* Vol. 375 (1981), pp. 90–122.

DeWitt, Bryce S. "Quantum Gravity." *Scientific American,* Vol. 249, No. 6, (Dec. 1983), pp. 112–29.

Denbich, K. G. *An Inventive Universe.* New York: Braziller, 1975.

Dermott, S. F., ed. *The Origin of the Solar System.* New York: Wiley, 1978.

Dicus, Duane A., et al. "The Future of the Universe." *Scientific American,* Vol. 248, No. 3 (March 1983), pp. 90–101.

Dodds, E. R. *The Greeks and the Irrational.* Berkeley: University of California Press, 1951.

Dott, Robert H., Jr., and Batten, Roger L. *Evolution of the Earth,* 3rd ed. New York: McGraw-Hill, 1981.

Drake, Stillman. *Galileo.* New York: Hill & Wang, 1980.

———. *Galileo at Work.* Chicago: University of Chicago Press, 1981.

———. *Galileo Studies.* Ann Arbor: University of Michigan Press, 1970.

Eddington, Arthur. *The Nature of the Physical World.* Ann Arbor: University of Michigan Press, 1958.

Eicher, Don L. *Geologic Time,* 2nd ed. Englewood Cliffs, N.J.: Prentice-Hall, 1976.

Eisley, Loren. *Darwin and the Mysterious Mr. X.* New York: Harcourt Brace Jovanovich, 1981.

———. *Darwin's Century.* Garden City, N.Y.: Anchor, 1961.

———. *The Firmament of Time.* New York: Atheneum, 1980.

Eliade Mircea. *The Myth of the Eternal Return.* Princeton: Princeton University Press, 1971.

Ferris, Timothy. *The Red Limit.* New York: Bantam, 1979.

Feynman, Richard. *The Character of Physical Law.* Cambridge, Mass.: MIT Press, 1965.

Fraisse, Paul. *The Psychology of Time.* New York: Harper & Row, 1963.

Fraser, J. T. *The Genesis and Evolution of Time.* Amherst: University of Massachusetts Press, 1982.

———. *Of Time, Passion and Knowledge.* New York: Braziller, 1975.

———, ed. *The Voices of Time.* New York: Braziller, 1966.

Gal-Or, Benjamin. "Are the Astrophysical and Statistical Schools of Irreversibility Compatible?" *Nature,* Vol. 234 (1971), pp. 217–18.

———, ed. *Modern Developments in Thermodynamics.* New York: Halsted Press, 1974.

Gale, Richard M., ed. *The Philosophy of Time.* London: Macmillan, 1968.

Galileo. *Dialogue Concerning the Two Chief World Systems.* Berkeley: University of California Press, 1967.

———. *Two New Sciences.* Madison: University of Wisconsin Press, 1974.

Gardner, Martin. *The Ambidextrous Universe.* New York: Scribner, 1979.

———. "Can Time Go Backward?" *Scientific American,* Vol. 309, No. 1 (Jan. 1967), pp. 98–108.

Geroch, Robert. *General Relativity from A to B.* Chicago: University of Chicago Press, 1978.

Gimpel, Jean. *The Medieval Machine.* Harmondsworth: Penguin, 1977.

Glass, Bentley; Temkin, Owsei; and Straus, William L., Jr., eds. *Forerunners of Darwin.* Baltimore: Johns Hopkins Press, 1959.

Gold, Thomas. "The Arrow of Time." *American Journal of Physics,* Vol. 30 (1962), pp. 403–10.

———, ed. *The Nature of Time.* Ithaca, N.Y.: Cornell University Press, 1967.

Goldstein, Thomas. *Dawn of Modern Science.* Boston: Houghton Mifflin, 1980.

Gott, J. Richard, III, et al. "An Unbound Universe?" *Astrophysical Journal,* Vol. 194 (1974), pp. 543–53.

Gould, Stephen Jay. *Ever Since Darwin.* New York: Norton, 1979.

———. "False Premise, Good Science." *Natural History,* Vol. 92, No. 10 (Oct. 1983), pp. 20–26.

Grant, Edward. *Physical Science in the Middle Ages.* New York: Wiley, 1971.

Gregory, Richard L. *Mind in Science.* Cambridge, England: Cambridge University Press, 1981.

Gribbin, John. *Timewarps.* New York: Delacorte, 1979.

Grisewood, H., ed. *Ideas and Beliefs of the Victorians.* New York: Dutton, 1966.

Guth, Alan H., and Sher, Marc. "The Impossibility of a Bouncing Universe." *Nature,* Vol. 302 (1983), pp. 505–6.

Hahm, David E. *The Origins of Stoic Cosmology.* Columbus: Ohio State University Press, 1977.

Hall, A. Rupert. *From Galileo to Newton.* New York: Dover, 1981.

———. *The Scientific Revolution,* 2nd ed. Boston: Beacon, 1966.

Hampson, Norman. *The Enlightenment.* Harmondsworth: Penguin, 1968.

Harman, P. M. *Energy, Force and Matter.* Cambridge, England: Cambridge University Press, 1982.

Harrison, Edward R. *Cosmology.* Cambridge, England: Cambridge University Press, 1981.

Hartline, Beverly Karplus. "Double Hubble, Age in Trouble." *Science,* Vol. 207 (1980), pp. 167–69.

Hartman, William K. *Astronomy: The Cosmic Journey.* Belmont, Calif.: Wadsworth, 1978.

Hawking, S. W., and Israel, W., eds. *General Relativity*. Cambridge, England: Cambridge University Press, 1979.

Himmelfarb, Gertrude. *Darwin and the Darwinian Revolution*. New York: Norton, 1962.

Holton, Gerald. *Thematic Origins of Scientific Thought*. Cambridge, Mass.: Harvard University Press, 1973.

Hoyle, Fred. *Ten Faces of the Universe*. San Francisco: Freeman, 1977.

Hubble, Edwin. *The Realm of the Nebulae*. New Haven: Yale University Press, 1936.

Huizinga, J. *The Waning of the Middle Ages*. Garden City, N.Y.: Anchor, 1954.

Irvine, William. *Apes, Angels and Victorians*. New York: McGraw-Hill, 1955.

James, William. *The Principles of Psychology*. Vol. 1. New York: Dover, 1950.

Kaufman, Walter. *Nietzsche,* 4th ed. Princeton: Princeton University Press, 1974.

Kaufmann, William J., III. *The Cosmic Frontiers of General Relativity*. Boston: Little, Brown, 1977.

Kazanas, Demosthenes; Schramm, David N.; and Hainebach, Ken. "A Consistent Age for the Universe." *Nature,* Vol. 274 (1978), pp. 672–73.

Kline, Morris. *Mathematics: The Loss of Certainty*. New York: Oxford University Press, 1980.

———. *Mathematics in Western Culture*. Oxford, England: Oxford University Press, 1980.

Kramer, Edna E. *The Nature and Growth of Modern Mathematics*. Princeton: Princeton University Press, 1982.

Krupp, E. C. *Echoes of the Ancient Skies*. New York: Harper & Row, 1983.

Landes, David S. *Revolution in Time*. Cambridge, Mass.: Harvard University Press, 1983.

Landsberg, P. T., ed. *The Enigma of Time*. Bristol, England: Adam Hilger, 1982.

Lansberg, Peter T., and Evans, David A. *Mathematical Cosmology*. Oxford, England: Oxford University Press, 1977.

Layzer, David. "The Arrow of Time." *Scientific American*, Vol. 233, No. 6 (Dec. 1975), pp. 56–67.

Le Goff, Jacques. *Time, Work and Culture in the Middle Ages*. Chicago: University of Chicago Press, 1982.

Lindberg, David C., ed. *Science in the Middle Ages*. Chicago: University of Chicago Press, 1978.

Lovejoy, Arthur O. *The Great Chain of Being*. Cambridge, Mass.: Harvard University Press, 1936.

MacDonald, D. K. C. *Faraday, Maxwell and Kelvin*. Garden City, N.Y.: Anchor, 1964.

Maddox, John. "Dispute over Scale of Universe." *Nature*, Vol. 307 (1984), p. 313.

Marcus Aurelius. *Meditations*. Harmondsworth, England: Penguin, 1964.

McMullin, Ernan, ed. *Galileo*. New York: Basic Books, 1967.

Meyerhoff, Hans. *Time in Literature*. Berkeley: University of California Press, 1955.

Misner, Charles W.; Thorne, Kip S.; and Wheeler, John Archibald. *Gravitation*. San Francisco: Freeman, 1973.

Morris, Richard. *Dismantling the Universe*. New York: Simon & Schuster, 1983.

———. *The Fate of the Universe.* New York: Playboy Press, 1982.

Mumford, Lewis. *Technics and Civilization.* New York: Harcourt Brace Jovanovich, 1963.

New Frontiers in Astronomy: Readings from Scientific American. San Francisco: Freeman, 1975.

Newman, James R., ed. *The World of Mathematics.* 4 vols. New York: Simon & Schuster, 1956.

Newton, Isaac. *Mathematical Principles.* Berkeley: University of California Press, 1946.

Nisbet, Robert. *History of the Idea of Progress.* New York: Basic Books, 1980.

Oppenheim, A. Leo. *Ancient Mesopotamia.* Chicago: University of Chicago Press, 1977.

Ornstein, Robert E. *The Psychology of Consciousness.* Harmondsworth, England: Penguin, 1975.

Page, Don N. "Inflation Does Not Explain Time Asymmetry." *Nature,* Vol. 304 (1983), pp. 39–41.

Park, David. *The Image of Eternity.* New York: New American Library, 1980.

Plato. *The Collected Dialogues.* Princeton: Princeton University Press, 1961.

Popper, Karl R. "Time's Arrow and Entropy." *Nature,* Vol. 207 (1965), pp. 233–34.

———. "Time's Arrow and Feeding on Negentropy." *Nature,* Vol. 213 (1967), p. 320.

Priestley, J. B. *Man and Time.* Garden City, N.Y.: Doubleday, 1964.

Prigogine, Ilya. *From Being to Becoming.* San Francisco: Freeman, 1980.

Quinones, Ricardo J. *The Renaissance Discovery of Time.* Cambridge, Mass.: Harvard University Press, 1972.

Raychaudhuri, A. K. *Theoretical Cosmology.* Oxford, England: Oxford University Press, 1979.

Reichenbach, Hans. *The Philosophy of Space and Time.* New York: Dover, 1957.

———. *The Rise of Scientific Philosophy.* Berkeley: University of California Press, 1951.

Robertson, J. Drummond. *The Evolution of Clockwork.* East Ardsley: S. R. Publishers, 1972.

Robinson, Arthur L. "New Test of Variable Gravitational Constant." *Science,* Vol. 222 (1983), pp. 1316–17.

Roux, Georges. *Ancient Iraq.* Harmondsworth, England: Penguin, 1966.

Ruse, Michael. *The Darwinian Revolution.* Chicago: University of Chicago Press, 1981.

Russell, Bertrand. *A History of Western Philosophy.* New York: Simon & Schuster, 1945.

Sachs, Robert G. "Can the Direction of Flow of Time Be Determined?" *Science,* Vol. 140 (1963), pp. 1284–90.

Sandage, Allan R. "Cosmology: A Search for Two Numbers." *Physics Today,* Vol. 23, No. 2 (1970), pp. 34–41.

Sandage, Allan, and Tammann, G. A. "The Hubble Constant as Derived from 21cm Linewidths." *Nature,* Vol. 307 (1984), pp. 326–29.

Segrè, Emilio. *From X-Rays to Quarks.* San Francisco: Freeman, 1980.

Shallis, Michael. *On Time.* New York: Schocken, 1983.

Shipman, Harry L. *Black Holes, Quasars and the Universe,* 2nd ed. Boston: Houghton Mifflin, 1980.

———. *The Restless Universe.* Boston: Houghton Mifflin, 1978.

Silk, Joseph. *The Big Bang.* San Francisco: Freeman, 1980.

Sklar, Lawrence. *Space, Time and Spacetime.* Berkeley: University of California Press, 1977.

Thomsen, Dietrick E. "The New Inflationary Nothing Universe." *Science News,* Vol. 123 (1983), pp. 108–9.

Toulmin, Stephen, and Goodfield, June. *The Discovery of Time.* Chicago: University of Chicago Press, 1982.

———. *The Fabric of the Heavens.* New York: Harper & Brothers, 1961.

Trefil, James. *The Moment of Creation.* New York: Scribner, 1983.

Tuchman, Barbara W. *A Distant Mirror.* New York: Knopf, 1978.

Turner, Michael S. "The End May Be Hastened by Magnetic Monopoles." *Nature,* Vol. 306 (1983), pp. 161–62.

von Weizäcker, Carl Friedrich. *The Unity of Nature.* New York: Farrar, Straus & Giroux, 1980.

Wald, Robert M. *Space, Time and Gravity.* Chicago: University of Chicago Press, 1977.

Waldrop, M. Mitchell. "Before the Beginning." *Science 84,* Vol. 5, No. 1 (Jan./Feb. 1984), pp. 44–51.

———. "Inflation and the Arrow of Time." *Science,* Vol. 219 (1983), p. 1416.

———. "The New Inflationary Universe." *Science,* Vol. 219 (1983), pp. 375–77.

Weber, Max. *The Protestant Ethic and the Spirit of Capitalism.* New York: Scribner, 1958.

Weinberg, Steven. *The First Three Minutes.* New York: Basic Books, 1977.

Whitehead, Alfred North. *Science and the Modern World.* New York: Free Press, 1967.

Whitrow, G. J. *The Natural Philosophy of Time,* 2nd ed. Oxford, England: Oxford University Press, 1980.

———. *What Is Time?* London: Thames and Hudson, 1972.

Wood, Charles T. *The Quest for Eternity.* Garden City, N.Y.: Anchor, 1971.

Woolf, Harry, ed. *Some Strangeness in the Proportion.* Reading, Mass.: Addison-Wesley, 1980.

Yahil, A. "The Density of the Universe from the Deceleration of Nearby Galaxies." *Annals of the New York Academy of Sciences,* Vol. 375 (1981), pp. 169–77.

Yourgrau, Wolfgang, and Breck, Allen D., eds. *Cosmology, History and Theology.* New York: Plenum, 1977.

Index